Practical Audio Amplifier Circuit Projects

Practical Audio Amplifier Circuit Projects

Andrew Singmin

Boston Oxford Auckland Johannesburg Melbourne New Delhi

Newnes is an imprint of Butterworth–Heinemann.

A member of the Reed Elsevier group

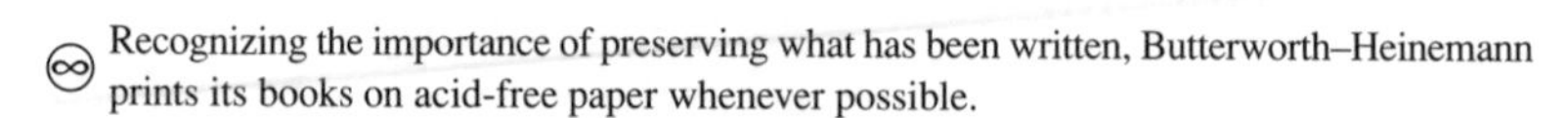

Recognizing the importance of preserving what has been written, Butterworth–Heinemann prints its books on acid-free paper whenever possible.

Butterworth–Heinemann supports the efforts of American Forests and the Global ReLeaf program in its campaign for the betterment of trees, forests, and our environment.

Library of Congress Cataloging-in-Publication Data

Singmin, Andrew, 1945–

Practical audio amplifier circuit projects / Andrew Singmin.

p. cm.

ISBN 0-7506-7149-1 (pbk. : alk. paper)

1. Electronics Amateurs' manuals. 2. Audio amplifiers Amateurs' manuals. I. Title.

TK9965.S545 1999

621.381—dc21 99-36689

CIP

British Library Cataloguing-in-Publication Data

A catalogue record for this book is available from the British Library.

The publisher offers special discounts on bulk orders of this book.
For information, please contact:

Manager of Special Sales
Butterworth–Heinemann
225 Wildwood Avenue
Woburn, MA 01801–2041
Tel: 781-904-2500
Fax: 781-904-2620

For information on all Butterworth–Heinemann publications available, contact our World Wide Web home page at: http://www.newnespress.com

10 9 8 7 6 5 4 3 2 1

Printed in the United States of America

Dedication

For SV, an adorable line dancer

Contents

Preface ix

1 **Introduction** 1

2 **Electronic Components** 7

3 **Building a Project Using a Basic LED Circuit** 29

4 **Audio Pre-amplifiers** 35

5 **Audio Power Amplifiers** 43

6 **Simple Filter Designs** 47

7 **Circuit Schematics** 55

8 **Test Instruments** 61

9 **Audio System Hookups** 71

10 **Construction Projects** 79

Project 1: Inverting Op-amp with Gain ×10 80

Project 2: Non-inverting Op-amp with Gain ×10 84

Project 3: Variable Gain Inverting Op-amp 88

Project 4: Inverting Buffer 92

Project 5: Non-inverting Buffer 95

Project 6: High Gain Inverting Amplifier with High-Frequency Cut Filter 98
Project 7: Pre-amp with Bass-Treble Control 102
Project 8: Power Amp with Gain and Bass Boost 107
Project 9: Audio Signal Generator 110
Project 10: Two-Input Mixer Pre-amplifier 114
Project 11: Pre-amplifier/Power Amplifier Combo 118
Project 12: Microphone Test Set 122
Project 13: Audio Test Set 127
Project 14: Guitar Pre-amp and Buffer 130
Project 15: Guitar Fuzz Pre-amplifier 134
Project 16: Electric Guitar Pacer 137

Index 141

Preface

In the course of writing this book I had the opportunity to research other electronics texts both in hard copy and on the now ubiquitous electronic websites, in particular searching out operational amplifier-based circuits. What I found (or more accurately, didn't find!) reminded me again of the inadequacy of texts with a practical focus, you know a book that is totally dedicated to the ins and outs of real practical circuits. There were too many texts and descriptions based on a theoretical aspect and hardly any (to tell you the truth I couldn't find any!) on how say to build a practical op-amp-based audio amplifier, be it pre-amplifier or power amplifier. All I found were the usual texts you'd find in manufacturer's data books (these are really just application notes). At best there might be a teaser circuit in the beginning and that's it. As I searched in vain for a more enlightened description, frustration began to set in, and the usefulness of producing a practical book started to take on a more spiritually rewarding dimension. If I found the frustration burdensome, then perhaps like-minded hobbyists have had the same experience.

My ideal kind of book (the type I'd go out and buy) is the type in which everything you are reading about can be built, and you know that the circuits are ones that have been already built and tested by the author. To give you an example; a real ac op-amp amplifier circuit requires a number of extremely necessary and critical components before it can function as a real circuit. As most of the published texts will invariably limit the circuit to just the gain setting components, that's all you get—so you then have to dig deeper and find out where the rest of the components go. For the hobbyist and especially the beginner, I would suggest staying away from application notes—although they may resemble real circuits, they're not. In addition, the application notes circuits are generally configured, that is, drawn, for dual-voltage supplies, and since hobbyists' circuits tend to most commonly function off a single 9-volt battery, that's another serious limiting factor to these circuits. Some of the electronics books that contain collections of electronic circuits unfortunately tend to include these.

The philosophy I've taken in this book, is to include only real working practical circuits. There are plenty of other textbooks that detail the internal workings of the op-amp. The vast majority of textbooks on the basic operating principles of the common op-amp, for example, will show you (a) circuits that are not practical working circuits, and (b) confusing component values.

Incidentally, rightly or wrongly, I seem to notice a correlation between how circuit schematics are drawn and whether they are real circuits. Real circuits tend to have a real solid line drawn for the power voltage rail and a real solid line drawn for the ground rail. Application notes tend to have arrows pointing to the power rail and arrows pointing to the ground rail. Interesting? My criteria for a practical circuit is, (a) I want to see a circuit that has real working values so that if I built the circuit exactly as shown, it would be guaranteed to work; and (b) I want the calculations to be simply and clearly stated. The vast majority of explanatory text I've come across for the op-amp are application note type texts; that is, they are theoretical descriptions of the workings of the op-amp. These serve no useful practical purpose as far as I'm concerned. As a hobbyist I want real practical information. This limitation in many of the available electronics texts is the main driving force for this book, and especially the way it has been written. The circuits described are 100 percent real! For example, if I'm building a pre-amplifier, there are just two things I need to know: (a) how do I control the gain of the amplifier, and (b) how do I actually get the circuit to really function. All my circuits show every single component needed to make them function correctly. There are critical components needed to get the circuit to work, and, sadly, most of the texts I've seen do not provide details of those components. This feature easily separates this book from what might on the surface appear to be similar texts.

An up-to-date listing of electronics terminology definitions can also be found in my recent publication, *Dictionary of Modern Electronics Technology*. This will be a useful reference to have.

Andrew Singmin
June 1999

CHAPTER 1

Introduction

You can find an abundance of topics on almost any conceivable aspect of electronics by scanning your local bookstores shelves or accessing any of the larger stores' websites. Practical books for electronics hobbyists are still in the minority, I believe, when stacked up against the proliferation of industrial electronics titles. Way back then in the 1950s, before the advent of the microprocessor, electronics projects of the home-brew variety enjoyed a following that has sadly fallen off in recent years. It costs significantly less to purchase an assembled electronic component compared to what it costs to build the unit from individual components. But if you are looking for an enjoyable hobby, and your aim is not to compete with commercially produced electronics, tinkering with components and coming up with circuits you can call your own can bring a lot of satisfaction. Many of the electronics books for hobbyists that I've come across over the years are inadequate in one way or another, marring the enjoyment of the projects that are featured in such books.

I've always felt very strongly about identifying what it is that is lacking in other similar volumes and trying to provide that information according to my preferences. Based on my past experience, some of the more obvious of these deficiencies are

1. The authors require that you use obscure and especially foreign components, particularly integrated circuits, that may be more widely available in another country (a country other than the one where the reader is located). These circuits aren't of the slightest value, since the components required to build the circuits cannot be readily obtained (in the United States).
2. The projects involve components, again especially integrated circuits, that are so specialized that procuring them as a noncommercial entity is almost impossible. These circuits generally provide

stunning features and capabilities, all through the magic of the custom application-specific integrated circuit. But what is the use of featuring these components if they are practically unobtainable by the general public? Even commercial companies sometimes have difficulties procuring these specialized devices unless they place a large order. Circuits based on such components are at best a curiosity—it's nice to drool over their features, but they are not really within our reach.

3. The projects call for radio frequency (RF) circuits that invariably require specialized coils and inductors. In my opinion, RF circuits have got to be some of the most difficult types of circuits to find the components for. I avoid like the plague any circuits that require specialized inductors. Incidentally, crystals also come in the RF category. Unless you can find a mail-order house specializing in these components, don't even consider the prospects of finding them. Trying to wind your own coils shouldn't even be contemplated.
4. Books may instruct you on building circuits that use supply voltages that are not supplied by standard batteries. We all know there are really only two regular battery voltage types you can buy at any corner drugstore: the 1.5-volt battery and the 9-volt battery. A 1.5-volt battery doesn't have enough power to drive much of anything, so the 9-volt battery is the only choice.

   Besides, some project cases also have a built-in battery compartment that accommodates a 9-volt battery. Since battery-powered circuits are safer than line-powered circuits and in any case have the benefit of portability, we really have only one choice for our power-supply voltage. Of course, commercial builders have access to other dc voltages such as 12 volts, but we're only concerned here with hobbyist circuits. If you're considering converting a circuit to 9 volts, this may or may not work; it all depends on how the circuit is configured. Trying to do this kind of conversion is not a good idea.
5. For some projects the circuits require a positive/negative supply, for example, a +9-volt and a –9-volt supply. You sometimes see two 9-volt batteries specified. This is totally unnecessary. It is cumbersome for a start; as far as I know, there are no project cases that accept twin 9-volt batteries, and you can get a plus/minus supply with a potential divider circuit. These circuits are likely to work with a single 9-volt battery whose voltage is split in half with a resistor network. The power output will be reduced because the voltage supply is halved. Personally I'm not a great fan of these dual-supply circuits.

6. Some of the circuits require a huge printed circuit board populated with dozens of ICs. I have found that any circuit requiring more than three or four ICs is not what I would describe as suitable for a beginner, because the greater the number of components, the more likely the circuit will be prone to errors during the build stage. Some of the "higher-level" electronics magazines tend to feature circuit designs populated with more components than you can count (or afford). Enough said about those circuits.
7. The projects may include circuits that contain really odd values for resistors and capacitors. In very few cases have I ever come across the need for what I would call "odd value" values for resistors or capacitors. When reading through the projects that follow, take note of the resistor values I use. In most cases, unless there's a really special need, all the circuits can be built using resistors of just a handful of common values such as 1 kohm, 10 kohms, and 100 kohms.

But enough about the negative aspects. You probably have a good idea of my pet peeves by now. What you'll find in this volume (and like, I hope) are thoroughly described circuits techniques and circuit projects, using readily available components, such as the LM 741 op-amp; a limited number of components values (to keep your budget down); designs that are simple and that have been built and tested before being published; and lots of useful tips as we go along.

The topic is audio amplifiers and at the end of this volume, if you follow the circuits as we go along, you'll know how and why these circuits are so constructed.

One of the most encouraging responses to be had from an electronics project that you had just completed is to see or hear it work! On that premise, no other type of circuit provides as much satisfaction as audio amplifier projects. Why? We are all very familiar with consumer audio appliances, from stereos to TVs, from CD players to Walkman cassette tape players. Sound is everywhere around us, be it CMT's country singers or blues musicians, we can identify with sound. It's a little more difficult to identify with the complexities of a microprocessor-based project, unless you are really computer literate. Anyone can listen to the sound ouptut from your first audio amplifier project and say, "That's a terrific sound—did you build that yourself?" I have always liked audio amplifier projects because they're so simple to put together and the result is both apparent and immediate. Hook up an electric guitar to your newly built guitar practice amplifier, and what do you get—instant sound! Hook up your microphone to a pre-amplifier/

power amplifier combination, and the effect is immediate. There's no ambiguity in deciding whether it works or not. If you can hear it, then it works!

The most common requirement in electronics has always been to amplify. Peruse through a variety of projects and you'll see the vast majority are either amplifiers or incorporate amplifiers within the design. Of the many types of amplifiers around, the ac (especially the audio amplifier) is, in my opinion, one of the most important because there is very little needed in the way of sophisticated equipment to bring the sound to life. You just need a speaker, and speakers are everywhere around us, for example in radios, TVs, and CD players. Later on as the chapters in this book progress, you will come to the core section of this book, the projects section. Here you will get a solid grounding into the way audio amplifiers work and find out how to build them for yourself. By working through a learning-by-building philosophy, you will soon be familiar with audio amplifiers by the time you reach the end of this book. The projects increment in complexity as we go along, but they are all easy to build. They've also been built and tested before being committed to the projects section.

Circuit Schematics

In electronics, a schematic is a way of representing the electrical connections needed to perform a certain function. Take a very simple case of a resistor potential divider used for continuously varying an input voltage, similar to the regular volume control found on every radio. The circuit schematic shows a potentiometer with three terminals. The two outer terminals are the fixed terminals, and the center terminal represents the center rotating wiper terminal. There's always a ground terminal (say terminal #1) for the current return connection, so that takes care of one of the fixed terminals (it is usually represented by the lower terminal). When viewed from the front, that is, with the potentiometer shaft facing you, the ground terminal is the right-hand terminal. Using that convention, the wiper voltage will increase as the potentiometer shaft is rotated clockwise. This is the customary way all volume controls function. The input voltage is connected to the other fixed left-hand terminal (terminal #2), and the output is taken from the center wiper terminal (terminal #3). When the potentiometer shaft is turned clockwise as far as it will go, the full input voltage is transferred to the output. What happens inside the potentiometer is that in the fully clockwise position, terminal #3 is connected to terminal #1 through the associated action of the rotating shaft. Conversely, at the fully counterclockwise position, the output voltage is zero. As the wiper is rotated clockwise, the voltage across the center terminal gradually increases. The output voltage continues to increase as the rotation continues, until finally the voltage output equals the voltage input. This can be easily seen if you have a potentiometer (anything around 10 kohms or 100 kohms in value), a 9-volt battery, and a voltmeter set to the voltage range.

My own introduction to schematics was no better than most beginners', I guess. Way back then, in the days of vacuum tubes, I had little if any idea what I was looking at. Having a good set of logically drawn and consistent schematics makes a world of difference to the beginner who is trying to figure out what that seemingly mysterious mass of lines and funny symbols must mean. So I can appreciate the benefit of having useful schematics to work with. If you're new to schematics, then work through the sections on that topic in Chapter 7, and you'll find it provides a good starting point.

Schematics describe an electrical function. Take your simple light switch operation. When the switch is turned on, the light comes on. To turn off the light, the switch is turned off. A schematic is used to represent this electrical operation. There are three components that will need to appear in this particular schematic: a switch, a light, and a power source. The connection mode is accurately depicted in the schematic; that is to say, the point-to-point wiring between the various electrical components is faithfully reproduced in our schematic. For this simple example the circuit schematic is basically a series connection between the power source, the switch, and the light. Whatever the complexity of the circuit, once the functional blocks have been defined, detailed electrical interconnections can be laid out.

For simple projects such as the ones shown in this book, a schematic can easily fit on a single sheet of paper. But for something as complex as a microprocessor, the schematics will cover a large area, and the complexity of the design will probably be understood only by designers in that field of discipline—but the principle is still the same. And in any case, any large schematic can always be broken down into smaller, more easily manageable blocks. Take, for example, a complex pre-amplifier and power amplifier schematic, perhaps like the one found in your high-powered stereo system. Just for fun, let's break it down into simpler blocks. First of all, it's a stereo system, and that means there are two identical blocks of everything, so get rid of one channel. The schematics are immediately half the size they were originally. Let's say we just wanted to look at the power amplifier section. Everything up to the volume control is confined to the pre-amplifier section, so get rid of that. The circuit is getting smaller. If you're wondering how you can know what to ignore, that judgment comes with experience, but the thought process I'm describing here can show you how to simplify a seemingly incomprehensive circuit.

Getting back to your stereo and the power amplifier. There will be a considerable number of stages leading up the final stage that drives the speaker. Each succeeding stage provides a little more amplification and enhances the signal to some extent. But, again, the basic function of each intermediate block is to amplify, so look carefully, and if it's an op-amp-based circuit (very likely), there'll be a string of integrated circuits. Focus way, way down to one of these blocks and you'll see the basic op-amp configuration. Lo and behold, you will probably recognize a few familiar circuit components (actually you'll have to work your way through this book before you can do that).

Knowledge through Understanding

Throughout this book I have tried to apply the philosophy that the best way to learn is to explain how things work as we go along, through the tests and circuit projects. My favorite example of how this mode of teaching is done is the explanation of how the positive input terminal of an ac operational amplifier is always set at a voltage bias point that is half the supply voltage. The lesson might sound a little wordy, but the implementation is very easy. If the supply voltage is a standard 9-volt battery, as it is for all of the circuits described here, then half that value is 4.5 volts. Actually, we don't even need to know what the supply voltage actually is; we just need to know that we should split the voltage in half. That can be done by connecting two resistors of equal value across the positive and ground terminals of the supply voltage. As an example, take a 9-volt battery and place two 100-kohm resistors that are connected in series across the battery terminals. With your multimeter set to the dc voltage range, measure the voltage between the resistor junction and the negative voltage terminal. It'll be half whatever the supply voltage is.

Rarely, if ever, will you see a similar explanation given when an ac audio amplifier circuit is described, and you'll wonder as I used to do, what the purpose was of those components feeding to the op-amp's positive input terminal. As far as we're concerned, it is sufficient to know that there are certain biasing conditions required. As to the reasons why that is so, that's another matter that has no effect on understanding the circuit design. With that example in mind, you'll see that all such circuits (that is, ac audio amplifiers) must (without exception) have those connections.

Suddenly you have the power of new knowledge. Open up any electronics hobby magazine, search for an ac op-amp-based design, and look for the biasing network. However convoluted the drawing of the network may be (and it often is convoluted, because of the idiosyncratic practice of having nonelectrical practitioners draw up electrical schematics), I guarantee that the biasing network will be there, in one form or another.

But enough of the introduction, because I'm sure you're itching to start the construction process. By all means flip to the projects section if you've already got my first volume, *Beginning Electronics Through Projects*, since you will have already got a good grounding from that book.

CHAPTER 2

Electronic Components

The components used in the projects section, Chapter 10, are given a brief introduction here. The descriptions of electronics components in this chapter are limited to those that are actually used in the projects in this book.

Components are the elements that breathe life into circuit schematics regardless of their complexity. Would you believe that the majority of the designs in this book are built using just four basic building block components? The four components that are the cornerstones of electronic designs are the resistor, the capacitor, the diode, and the transistor. What about the integrated circuit? I hear you ask. Integrated circuits are collections of usually just these four elements. We'll consider ICs along with other important components later in the chapter.

Building Block Components

Resistors

The humble resistor is by far the most prolific component in use, so it makes a good starting point. A resistor, as the name implies, serves to provide some form of resistance, which is measured in ohms. Even the very name *resistor* already presents an inkling of what it does. In its very simplest form, as a stand-alone component, a resistor presents a resistance to the current flow that would normally take place when voltage is applied to a circuit. A high resistance presents more of an "obstacle," so the resulting current flow is relatively small. On the other hand, a low resistance allows more current to flow. If a resistor were connected in series with a current source it would be acting as a current limiter. With resistors you can carry out a lot of simple experiments that are easy to understand and explain. For instance, put a resistance in series with a voltage source and a lightbulb: as the resistance goes up, the light dims, and as the resistance goes down, the light brightens. What could be easier to understand?

If limiting current flow through a circuit were all there is to a resistor's function, then we wouldn't have much of a range of circuits to play with. But human ingenuity being what it is, we (electronics designers) have a lot more uses for the resistor. What can we do with two resistors? As you will soon see, the ingenuity or cleverness of the application is tied into the situation in which the resistor is being put to use.

The humble light emitting diode's (LED's) sole function is usually no more than to produce light and to serve as a solid-state indicator lamp. Driven from a low-voltage source, the LED nevertheless has to have a current-limiting resistor inserted in series with the voltage source. Where else do we find the innocent resistor lurking? Operational amplifiers, or op-amps, have a devastating amount of power packed into a tiny 8-pin dual-in-line (DIL) package. Gain setting, the most common feature for an op-amp, is determined by two resistors. Regardless of the sophistication and variety of op-amps (and there are many), they all have to depend on the lowly resistor to function. A resistor is like the mortar holding the bricks together that ultimately form a house. Mortar's not much to look at or get excited about, but where would bricks be without it?

Split bias voltages are found everywhere in op-amp circuits running off a single battery. The positive non-inverting pin must be biased in order to halve the supply voltage. Two resistors of equal value placed across the supply voltage and ground nicely provide the required split voltage.

In a slightly different form, but nevertheless still a resistor, there is the potentiometer, which is nothing more than a variable resistor. Figure 2-1 shows the two basic resistor types. All radio receivers, stereo amplifiers, cassette recorders, and other such devices have volume controls for obvious reasons. Resistors come in a variety of practically infinite values, from the typically used values of a few ohms to a few megohms. The LED example uses a current-limiting resistor that can vary from a few hundred ohms to a few thousand ohms depending on the supply voltage and the LED brightness required. Gain-setting resistors can range anywhere between a few kohms to

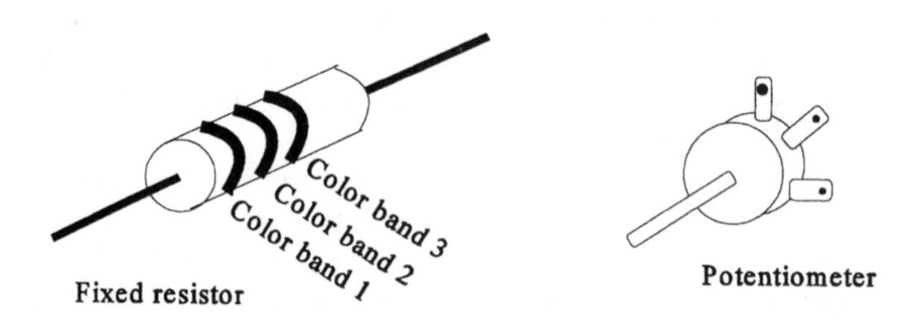

FIGURE 2-1 Fixed resistor and potentiometer

a Mohm. Resistors for the split bias supply typically are 100 kohms in value. Resistors are usually associated with dc circuits, as we've seen, and provide a number useful functions, but most commonly they control current.

Other than limiting current, one of the next most common functions of the resistor is to act as a potential divider circuit. In the simplest case, two equal resistors are placed across a simple voltage source (for example, a 9-volt battery). The resistor midpoint is half the source voltage, that is, 4.5 volts. This can be checked with a multimeter set to measure dc volts. Ohm's Law tells us we can find the current flow in a resistor by dividing the applied voltage by the resistance. The voltage source for the projects in this book is always a 9-volt battery. For the ease of the arithmetic I just round this up to 10 volts. So if we've got a 10-ohm resistor, the current is just under 1 milliampere (mA), actually, it is 0.9 mA, as the current is the ratio of the voltage to the resistance. That quick calculation gives us an idea of what to expect for our meter reading.

The same multimeter set to the ohms or resistance range can be used to check out resistor values. There are two precautions if you're going to do this now: Keep your fingers away from the resistor terminals because your body has a finite resistance, more if your hands are sweaty and less if they're dry. What you're doing when you touch the resistor terminals is adding your body resistance to that of the resistor you're trying to measure. The other precaution is to zero the resistance meter first. Do this by shorting the meter terminals and adjusting the "zero knob" until the meter reads zero. You need only do this with the analog type of multimeter.

The value of a particular resistance is marked on the component body, typically with a three-color band code. A fourth band represents the tolerance, but for the sake of simplicity you may ignore this if you just want to read off the resistor value (which is generally the case). As you almost certainly will want to be able to read resistor color codes, here they are:

Color Band	Equivalent Number Code
Black	0
Brown	1
Red	2
Orange	3
Yellow	4
Green	5
Blue	6
Violet	7
Gray	8
White	9

Often people will make up their own jingle to remember the color codes—you know, something that has meaning for you, such as **B**ye **B**ye **R**eba **O**ff **Y**ou **G**o **B**e **V**aliant **G**o **W**ell. You get the idea.

The next most common format for resistors, and one that you'll come across very often in the circuit projects, is the variable resistor, or, as it is more usually called, the *potentiometer.* Relatively speaking, the potentiometer is a much larger device than the resistor; it is more mechanical as opposed to electrical, and it is a three-terminal device. A rotating shaft coupled internally to a movable wiper track follows an arc-shaped path over a track of resistive material. The movable wiper terminal is brought out to a fixed electrical connection point. Further, two fixed terminals are electrically connected to the other two ends of the resistive track. As you can probably tell, the resistance measured across the wiper terminal and either of the other ends will vary continuously as the shaft is rotated. The maximum resistance value will be the value marked on the device; typically values of 1 kohm, 10 kohms, and 100 kohms are used.

Resistor values will typically run from 1 ohm to 1 Mohm. I find that with most circuit applications you can get away with using just a few "good" resistor values. My own personal preference is 10 ohms, 100 ohms, 470 ohms, 1 kohm, 2.7 kohms, 4.7 kohms, 10 kohms, 27 kohms, 47 kohms, 100 kohms, 470 kohms, and 1 Mohm. These values can be further distilled down to 100 ohms, 1 kohm, 10 kohms, and 100 kohms if I had to choose the four most useful values. Look at the circuits later in the book and see how often these values turn up. Intermediate values can be built up by juggling a handful of basic values and learning a bit of "resistor math." Two resistors of equal value connected in parallel produce half the resistor value. So two 1-kohm resistors produce 500 ohms, and two 10-kohm resistors give you 5 kohms. So if a circuit called for a 5.5-kohm resistor and it's late at night, and you desperately need that last component to finish, join two 1-kohm resistors connected in parallel to two 10-kohm resistors connected in parallel, and you've got what you need. A useful trick indeed.

The more general rule to follow when the resistors are not equal in value is that for two resistors of unequal value connected in parallel, the total value is the product divided by the sum of the two values. For example, a 1-kohm and a 10-kohm resistor connected in parallel will yield the product $10 \times 1 = 1$, divided by the sum of the resistor values, $10 + 1 = 11$, yields $10 \times 11 = 0.9$ kohm. Another useful trick to remember when connecting two resistors in parallel is that the total is always less than the smaller of the two values. In the example above, 0.9 kohm is less then 1 kohm (the smaller). For more than two resistors connected in parallel (you can use as many resistors as you want), the rule is

1/total resistance=1/resistor 1 + 1/resistor 2 + 1/resistor 3

Here's another example. A 1-ohm, a 2-ohm, and a 3-ohm resistor are connected in parallel. The result is

1/total resistance=1/1 + 1/2 + 1/3 = 1 + 0.5 + 0.33 = 1.833 ohms

To check our math, since 1/total resistance is 1.833, the total resistance is 1/1.833 = 0.545 ohm, and this value is less than the smallest value (1 ohm). On the other hand, adding two or more resistors in series (end to end) merely gives you the sum of all the individual resistor values. A 1-kohm resistor and a 100-kohm resistor connected in series thus yields 101 kohms. So by combining resistors in series and parallel you could make up almost any value you want. Figure 2-2 shows the series, parallel combination. But it's much easier to go out and buy a resistor with the value you want (and that one resistor will take up less space).

Apart from the actual resistance value, there is a second parameter associated with resistors, the tolerance rating, and it is designated by an extra color band. The most commonly specified tolerance is 5 percent (a gold band), followed by 10 percent tolerance (indicated with a silver band). In case you encounter them, there are also resistors with no color band that are equal to 20 percent tolerance, but it is inadvisable to use them because they tend not to be accurate. The tolerance percentage refers to the spread of values on either side of the nominally marked value (the three color bands) that the resistor is allowed to read and still remain within specification. This tolerance designation gives the resistor manufacturer a

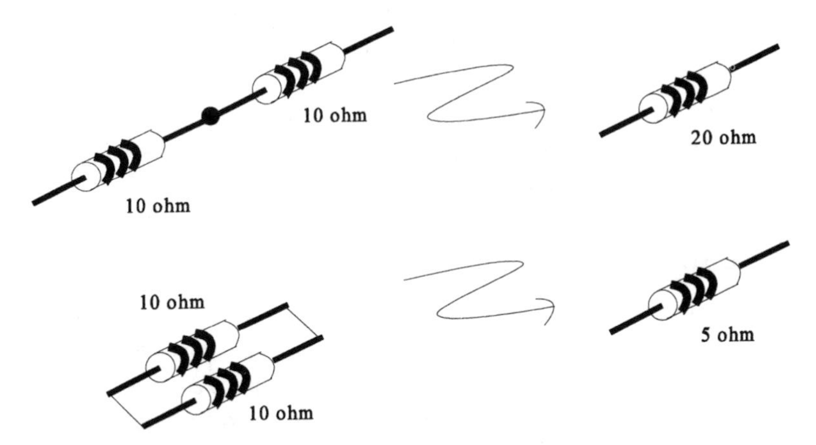

FIGURE 2-2 Resistors in series and resistors in parallel

greater latitude in offering resistors with a nominal value than would be otherwise possible. From the user's point of view (you and me), this means a 100 kohm resistor might not exactly read that value when measured, but it is perfectly acceptable from the manufacturer's point of view. For example, if you have a 5 percent 100-kohm resistor and you measure the actual resistance, it could lie anywhere between

100 kohms + 5 percent = 100 kohms + 5 kohms = 105 kohms, or
100 kohms – 5 percent = 100 kohms – 5 kohms = 95 kohms.

If this were a 20 percent 100-kohm resistor, then the limits would run from 120 kohms to 80 kohms—which is an extraordinarily wide variation. All the projects described later in the book, in Chapter 10, use 5 percent tolerance resistors.

The third parameter associated with resistors is their power rating. The value typically used is 1/4 watt, which is also the wattage specified for the project circuits in this book. The power rating of a resistor refers to its ability to dissipate power, which in turn translates to its ability to dissipate heat. The more current you pass through a resistor, the hotter it gets, and the resistor power rating must be sufficient to stand up to the dissipated power. Larger resistors go up to1/2 watt and more. It's a waste to use these for the projects in this book because these resistors take up more space, cost more, and are unnecessary. But for the sake of demonstrating the calcuations involved, I'll describe what happens to the power rating when we join resistors in series or parallel. In the simple case of two 100-ohm 1/4 watt resistors joined in series, the total resistance is 200 ohms, and the power rating is still 1/4 watt. But when these resistors are joined in parallel, the resistance drops to 50 ohms, and the power rating increases to1/2 watt—a nice technique to remember if you want to increase your power rating. Let's say you wanted a 10-ohm 1-watt resistor and the shops are closed. This is quite a large beast. You've got a bunch of common 100-ohm 1/4-watt resistors. Take ten of these 100-ohm resistors and connect them in parallel. The total resistance is now 10 ohms (one tenth of the individual values), and the power is increased to $10 \times 1/4 = 1.25$ watt. This is another good trick to remember.

Capacitors

Capacitors, like resistors, are two-terminal devices and are distinctive in terms of their ability to block dc signals and pass ac signals. For example, a dc signal, that is, voltage from a battery, cannot be passed through a capacitor, but an ac signal, say it's coming from transistor radio's earpiece socket, will pass through a capacitor. A resistor, by comparison, will pass both ac and dc signals by the same amount.

In practical circuit situations there are many instances in which the ac signal has to be passed but the dc component needs to be blocked. One such instance is when a power amplifier's signal is fed to a speaker. You'll always see a capacitor feeding the signal to the speaker. Another area in which you'll always notice the presence of capacitors is at the input and output of ac amplifiers. Capacitors are measured in units of farads, but since these are very large units, the much smaller units of pico-, nano- and microfarads are most often used. A picofarad is 10^{-12} farads, a nanofarad is 10^{-9} farads, and a microfarad is 10^{-6} farads. The conversion between the units is such that 1 pF equals 10^{-6} µF.

Remember the simple LED circuit we discussed, the one with the resistor acting as the current limiting device? If the resistor were replaced by a capacitor, the LED would not function because no dc current would be allowed to pass through. Capacitors have a property that is equivalent to dc resistance; they have ac resistance or reactance. The capacitor's reactance is calculated in ohms (like that of the resistor), and it is a function of the frequency of the signal under consideration. The capacitive reactance is inversely proportional to frequency; that is, as the frequency increases, the reactance decreases.

Capacitors can be broken into two basic categories based on their physical structure: the simple non-polarized type, which is also small in size and small in electrical value (that is, capacitance); and the larger polarized type, with higher associated capacitance values. Figure 2-3 shows the two basic types. Figure 2-4 shows the series, parallel combinations. Figure 2-5 shows the axial, radial types.

Capacitors such as the electrolytic capacitor are polarity sensitive, which means that they have to be connected in a certain way in the circuit. The electrolytic capacitor is a polarized component, and markings on the body of this capacitor indicate the appropriate negative and positive terminals. As a general rule, capacitors above and including 1 µF in value are usually polarized. Capacitance values for the components with larger values are

FIGURE 2-3 Disc and electrolytic capacitor

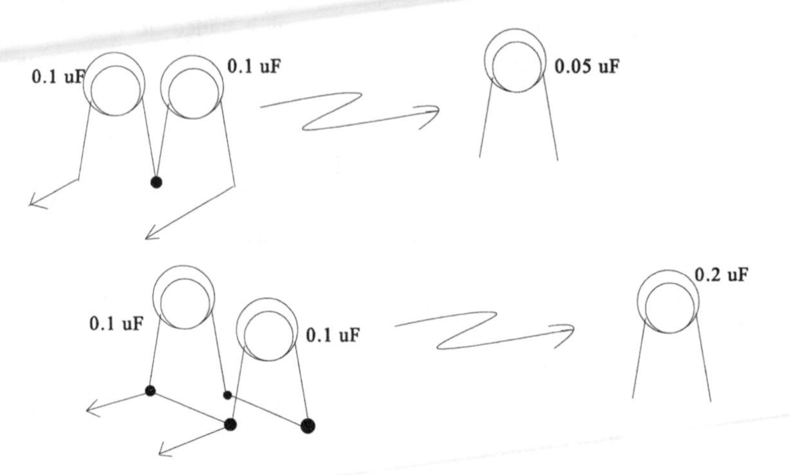

FIGURE 2-4 Capacitors in series and capacitors in parallel

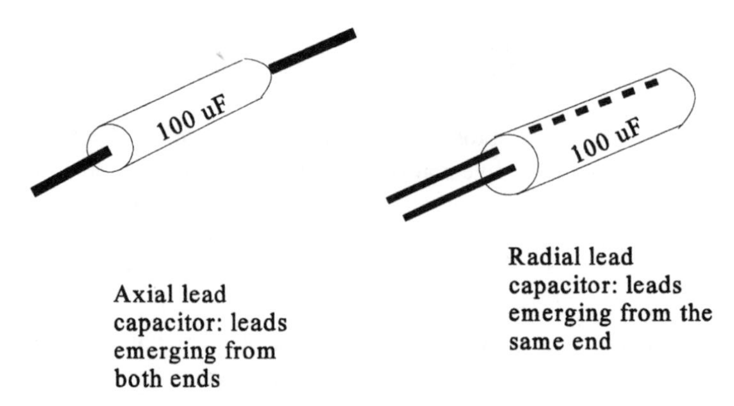

FIGURE 2-5 Axial and radial capacitor types

marked on the component's body since there is sufficient space to print out the value in full; that is, 1 μF will actually be printed on the body of the capacitor. The values of capacitors with smaller values are represented with a unique numbering code. The system is similar to the color coding used for resistors, except numbers are used instead of colors. There are three numbers to represent capacitance. It's much easier to understand the system by way of an example. Let's look at the code 104. This is a capacitance value expressed in picofarads. The first and second numbers relate to the actual first two digits of capacitance. The final number indicates the number of zeros following. So 104 is 100,000 picofarads. Since this number is a bit unwieldy, multiply it by 10^{-6} to convert to μF, and this works out to 0.1 μF, a much more convenient number to work with. This is a very common capacitor value as you'll learn later when you read about the projects in Chapter 10.

Variable capacitors do exist, but they are used less frequently than are variable resistors. But variable capacitors are still two-terminal devices. Why? Variable capacitors operate on the principle of varying the overlap between two metal plates, separated by either air or an insulator—the greater the overlap, the greater the capacitance. So you see, just two terminals are needed. There are no variable capacitors used in the projects in this book.

Radial lead capacitors have leads emerging from one side of the body, and if you don't have any height restrictions in your project case, this is the type I recommend you use. Axial lead capacitors, on the other hand, have leads emerging one from each end of the body of the component. They take up an awful lot of board space and are used only when the assembly board profile has to be as low as possible, but this is hardly a requirement for simple single-IC hobby projects. (An example of a requirement where you would need a very low profile, would be for a pager. Pagers are thin as we know and therefore, need an assembly board with a low profile.)

Like resistors, capacitors can also be connected in series and in parallel to form different values. But the rules are different from those for resistors. To increase a capacitor value, we connect two together in parallel. So two 0.1 μF capacitors connected in parallel gives us 0.2 μF. Three capacitors of 0.1 μF value each connected in parallel gives us 0.3 μF, and so on. If the capacitors were to be connected in series, then

1/ total capacitance = 1/capacitance 1 + 1/capacitance 2, and so on.

For example, two 0.1 μF capacitors connected in series results in a 0.05 μF capacitor, since

$$1/\text{total capacitance} = 1/0.1\ \mu\text{F} + 1/0.1\ \mu\text{F} = 10 + 10 = 20$$

Hence, the capacitance is 1/20 = 0.05 μF. Sometimes for timing applications in an oscillator circuit, you might want to change the output frequency a little, and this is one way of obtaining a 0.05 or 0.2 μF capacitor if you don't have one handy (and it's too late to run out to your local component store).

For ac applications, an approximate counterpart to the resistor is the capacitor. Again, a seemingly innocent two-terminal device, the capacitor appears lowly in form, but it is critically needed, like the resistor. Consider any amplifier circuit as an example.

Returning to the ac amplifier example, we can see later when examining these circuits in Chapter 10, that there is always a capacitor coupling the signal in and coupling the signal out. That's the way to recognize an ac amplifier by the presence of the capacitor at the input and the output. For simple pre-amplifiers, the coupling capacitors, as they're called, could be around 0.1 µF in value. If we assume the signal to be in the audio frequency range, say 10 kHz, then the capacitive reactance works out to be 159 ohms. This is very low and practically a short circuit. As the capacitive reactance scales inversely with the capacitance, doubling the capacitor to 0.2 µF will reduce the capacitive reactance by half to 79.5 ohms. In our example of the split supply with the resistor we saw that two resistors of equal value gave us the split voltage. Generally in an actual working circuit, you will see a capacitor placed across the lower resistor, that is, the one connected to ground. This is typically a capacity with a large value (100 µF), which is really a short circuit at the audio frequencies we are working with. Another very common way of connecting a capacitor is directly across the supply line, that is, between the plus and minus voltage rail. With a battery supply this is not so critical, but if you're using a low-voltage line adapter, using a large value smoothing capacitor (several 1000 µF in value) will aid in producing a smoother supply source.

Diodes

Diodes are two-terminal devices that have a feature that is totally distinct from the features of resistors or capacitors. They are distinctly polarity sensitive. When dc voltage is applied to a diode, a high current will flow in one direction, but reversing the voltage will to all intents and purposes cause no current to flow. Or put another way, when the diode is configured in what is called the forward-biased mode, the diode will conduct current. Reverse the bias to the reverse-biased mode and no current will flow. This is defined as a rectifying action. Ac voltage, say originating from the line voltage, can be immediately converted into a dc voltage of sorts by feeding it through a diode. The diode essentially passes on only half of the positive and negative going waveform. Electronic circuits are invariably powered with the positive voltage supplying the power rail.

To test this out, connect up a resistor, say 100 kohms, across a 9-volt battery with a current meter inserted between the positive battery terminal and one resistor terminal. Make sure the current meter's positive terminal goes to the battery's positive terminal. The current will be just under a tenth of a milliamp. The actual current value doesn't matter. If the meter's needle kicks against the end stop, reverse the meter polarity (assuming you've got an analog multimeter); a digital multimeter will automatically compensate whatever polarity is present. When using a digital multimeter to measure dc voltage, there is no need to worry if you get the test leads reversed. The mul-

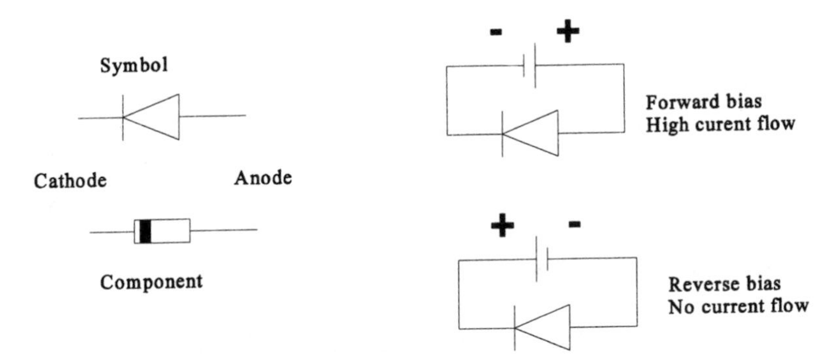

FIGURE 2-6 Diode symbol and bias conditions

timeter will still show the correct voltage; there's just a negative sign in front of the numeral. That tells you that the multimeter red test lead, for example, has been connected to the negative voltage potential. There is no damage done to the digital multimeter. If you now take the feed of the positive battery terminal via a diode (it doesn't matter at this stage which way round it goes), one of two things will happen. Either the current will be the same as before, or the current will be zero. Whatever it is, take note of it. Then reverse the diode polarity; just reverse the diode's connection in the circuit. An effect opposite to the one you first observed will now take place. You're seeing the rectifying action of the diode.

One really useful function for the diode is as a protective device. Electronic circuits are invariably powered with the positive voltage supplying the power rail. If the voltage is inadvertently reversed, there is a high probability that the components will suffer some damage. Placing a diode (this would be a power type called a rectifier) in series with the positive supply voltage would do the trick. When the polarity is correct, insert the diode in such a way that current starts to flow (trial and error is the quickest way to learn which way to attach the diode, if you're not sure about the markings). Now, if the voltage polarity should be reversed, no current will flow, thus providing the protection. Try it and see.

Transistors

Transistors are totally different from resistors, capacitors, and diodes. The latter are what are termed passive components, performing a singular function as we've seen, useful certainly, but not active in the electronic sense. A transistor is a truly active device. It can take a signal and amplify it. A number of support components are needed to make the transistor into a working amplifier—you guessed it, using a few resistors and capacitors again. Depending on the designer's talent, transistors can be configured into an endless string of circuits, amplifiers, oscillators, filters, alarms, receivers, transmitters, and so on. The versatility of transistors knows no bounds.

Although I do not include transistor-based circuits in this book—the reason being that integrated circuit projects are so much more well behaved and, hence, simpler to design—I do provide a brief overview on transistors, since integrated circuits are really just a huge collection of transistor-based circuits. Transistors are three-terminal devices; the terminals are known as the emitter, the base, and the collector. Figure 2-7 shows transistor details. Transistors come in two "flavors" so to speak: the more common NPN type operates with a positive supply voltage, and hence, it is very compatible with integrated circuits, which almost always run on a positive supply. The less common transistor type is the PNP device, and that, as you might have guessed, requires a negative supply voltage (not so commonly found in circuits).

Transistors are defined as active devices because they have the capability, given the appropriate support components, to perform useful functions; the most common of these is amplification, but the other is oscillation. A simple, common emitter amplifier can be designed around four resistors and a capacitor as well as the usual input and output coupling capacitors. But there are two main reasons to use the integrated circuit (IC). The amplifier's performance is influenced by the transistor's parameters, not so with the IC. Coupling the transistor amplifier into a following stage requires careful consideration of the loading effect. An IC-based amplifier just gets coupled into the next. The IC amplifier is such an effortless pleasure to use. The input, output, and gain are so nicely controlled. You would have had to have labored through the transistor's design quirks to really appreciate how much more controlled the IC is.

Transistors come in a huge variety of types, from general-purpose, small signal (the most common) to large power devices. The frequency range of operation can extend from dc to audio all the way up into the microwave range. Transistors are not as easy to evaluate as ICs. Put together a few resis-

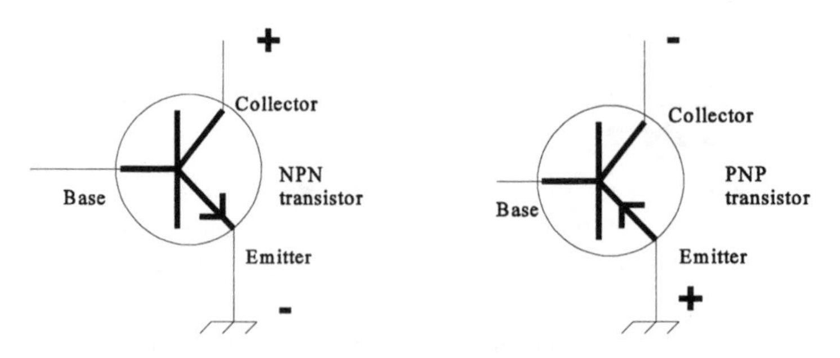

FIGURE 2-7 Transistor terminals

tors and capacitors around an IC and you'll soon know if the circuit's working (and it usually is), since you don't have to even wonder if the IC itself is working. But try the same with a transistor, and you'll find that determining whether or not the circuit is working is a lot harder. Was the transistor the right type? Was the bias network correct? Is the circuit design right? If the transistor circuit doesn't work, you'll always wonder whether the transistor itself is OK for the application. Isn't it great to know that in the majority of cases, you need only ask for one IC (the LM 741 as it turns out) when working with ICs. Enough said about transistors. They have their uses in specific applications, but you've got of be a bit more circuit smart.

Other Components

Integrated Circuits

The integrated circuit is an amazingly robust bullet-proof device, by which I mean that you can put practically any design around the IC and know that it is going to behave itself–OK, behave itself within reason, but ICs are brilliantly transparent compared to transistors. A small handful of resistors and capacitors and, hey, presto, we've got a well-behaved amplifier. The transistor could never match that! I know the comparison's a little unfair, especially because the IC itself is composed of a very carefully designed collection of transistor-based circuits, but we're talking user-friendliness here. I recall the difficulty I experienced way back in the mid-1960s getting a simple half-watt transistor power amplifier to function properly. The component count was high, special parts were difficult to come by, set up was tricky, and current consumption was high. Now we've got the LM 386 audio power amplifier on a chip! It's actually been around for a considerable number of years, but it is still very widely used. One IC and two capacitors and you're in business—Wow! and the current consumption's pretty good, too. The LM 386 IC is an example of a special function IC that is designed to deliver (which it does admirably) just one unique function. The unique one function for the LM 386 IC is as an audio power amplifier. It's hard to believe that a small 8-pin plastic part, little bigger than one of the buttons on your TV remote, packs such a technological punch. This particular IC runs nicely off a regular 9-volt battery—there are no weird dual supplies to worry about. Many of other higher power ICs require dual supplies, or voltages of 12 volts and higher (a 12-volt battery that you can't buy off the shelf and that would fit in a project case), and consume masses of current.

Integrated circuits fall into two broad categories; analog and digital. They are very easily recognized in terms of their functionality and also in terms of the way they're depicted in circuit schematics. Analog ICs process mostly ac signals, but they also process dc signals. The absence of a coupling capacitor at the input would signify that this is a dc amplifier we're looking at. A dc amplifier has to be capable of amplifying both dc signals as well as ac signals. Analog signals, such as audio signals, require coupling capacitors at the input and output because only ac signals are allowed to be coupled through the amplifier. The presence of the coupling capacitors removes the dc components. The schematic is also drawn in the form of a sideways triangle representing the IC. Input goes into the wide end on the left and exits as an output from the pointed end on the right. In essence all analog IC blocks resemble this basic form. Typical examples of analog ICs are the LM 741 general purpose op-amp in an 8-pin DIL package and the LM 324 quad op-amp package in a 14-pin DIL package. When space is at a premium, the LM 324 is a superb device; it is especially suited for audio applications, and it occupies far less board space than do four separate LM 741s. Analog ICs, incidentally, are also called linear ICs. Digital ICs only use two voltage states, a logic high (1) and a logic low (0). There are no capacitors in the signal coupling lines, and the schematics are generally drawn in the shape of rectangles or squares. Typical examples can be found in the 7400 series of digital TTL ICs. There are no digital ICs used in this book, but it's worthwhile to make a quick mention of them here because they're such a major portion of the IC family.

The third group of ICs that we cover in this book are special function ICs, that is, devices falling into neither the analog nor the digital category. Analog or digital ICs don't really do anything by themselves, so to speak. To turn an LM 741 into an amplifier (which is usually the case), you have to adjust the rest of the circuitry. Alternately the '741 could be designated as an oscillator, and again it is changed accordingly. Digital ICs operate on the principle of responding to just two voltage levels, a low level (also called a '0') and a high level (also called a '1'), and hence, are also called logic devices. Digital ICs can be thought of as a series of logic gates that are configured to perform a certain logic function. Special function ICs are complete in themselves. The LM 386 audio power amplifier that we'll be focusing on heavily later in the book is just that; it is a self-contained unit that is designed to perform just one task (and it does so admirably at that). Another much used special function IC is the LM 555, timer IC, so commonly used to provide a train of square wave pulses. Figure 2-8 shows the basic IC outline.

Switches

Switches occur in so many places in spite of their somewhat mundane nature. After all, a switch is just an on/off device. There are actually many different configurations for switches, and it's a good idea to get to know the

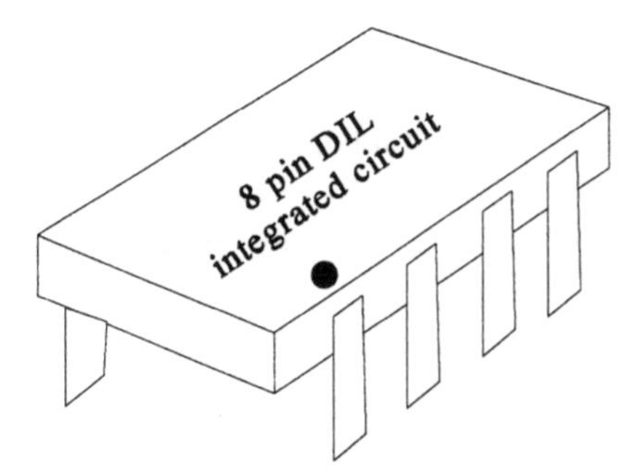

FIGURE 2-8 Integrated circuit package outline

variations. First of all, there's a terminology specific to switches: *poles* and *throws*. The simplest type of switch, like the type you'd find in a lamp switch, is called a *single pole, single throw* or *SPST* switch. The simple SPST switch has two terminals, one of these goes to the source (this being typically the positive voltage supply from a battery) and the other goes to the output (typically this would be the circuit that is to receive the power from the battery), hence, the output can only be connected to one terminal. It also has a toggle that flips back and forth. The light flips on one way and off the other. Switches always have to be described with respect to an input signal and an output signal. The pole refers to the number of terminals the input can be connected to. With the SPST switch there is just one. The throw refers to the number of terminals the output can be connected to. In the SPST switch there is just one. What if we had two terminals to which the output could be connected? There are now two throws, so this kind of switch is called a *single pole, double throw switch*. In this switch there are actually three terminals arranged in a row. The input attaches to the center terminal, and the other two terminals go to the two outputs. The SPST switch, as we've seen, is the type used to switch an appliance on and off. The SPDT can be used to switch either one of two lights on. This kind of switch is not too useful in real life, since there is a chance you may want both lights off. But it illustrates the point. Incidentally, there is a less common type of enhanced version of the SPDT switch with a center off position. The toggle is biased mechanically so it can be positioned in between the two extreme positions. That switch will turn off either lights (in our example). In the above example we have had the switch connected just in the positive supply line (where it is usually connected). The other terminal, that is, the negative terminal, if we were considering, say, a battery being hooked up to a light, would be permanently connected into the circuit. In situations where both sides of the battery need to be switched, we use a switch that is essentially a dual version of the SPST switch. This switch has two sets of terminals, each set identical to the other in function. As you might have guessed, this is a *double pole, single pole,* or *DPST,* switch, where a pair of inputs can be switched to a pair of outputs. This switch type is useful because it makes possible more than just the basic on/off function. An even more versatile switch is the *double pole, double throw,* or *DPDT,* switch, where two separate inputs can be switched to two separate pairs of outputs.

Table 2-1 below illustrates the use of the different switch types. Figures 2-9, 2-10 depict the switch types very clearly.

The dotted line for the DPST and DPDT switches indicate that these switches have ganged contacts; that is, they are switched together with each mechanical toggle. For a seemingly simple mechanical device, there's certainly more to the humble switch than you first thought. Apart from the switching differences, switches also come in different current ratings; the higher the current capacity, the larger the physical switch. For the circuit projects shown in this book, choose switches with the smallest current ratings available. Here's a very useful tip, I only found through experience: some small switches (the toggle type) require a huge amount of force to toggle between positions. What this means is that if you've got a very light plastic project case with this type of switch mounted on the front panel, you will

TABLE 2-1 The Uses of Different Switch Types

Switch Type	*Purpose*
SPST	Used to switch a single mono amplifier speaker on or off
SPDT	Used to switch a mono amplifier between two speakers
DPST	Used to switch a single pair of stereo amplifier speakers on or off
DPDT	Used to switch a stereo amplifier between two pairs of stereo speakers

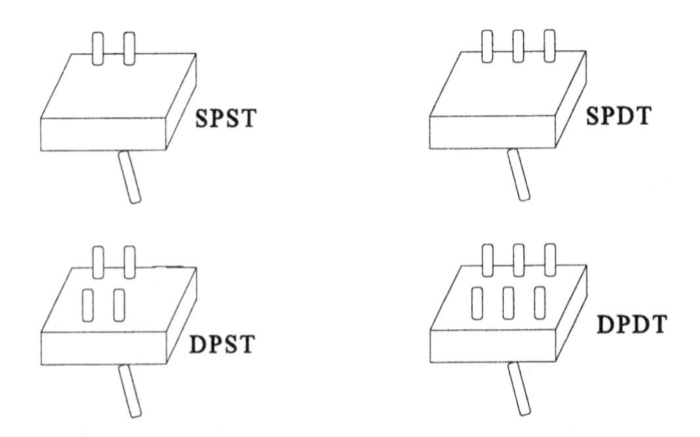

FIGURE 2-9 Switch terminals

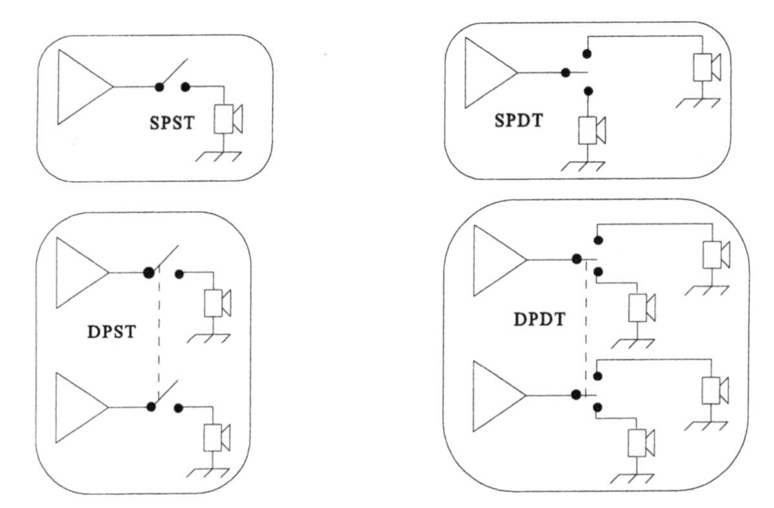

FIGURE 2-10 Different switch applications

most likely tip over the case when you try to flip the switch. I found this out the hard way! So choose small switches that have a very light toggle action. A slight flick of your finger should flip the switch to the other position. Switches are quite costly, and you can save yourself a bundle by not buying the wrong type.

Rotary switches are like super versions of the regular switch, and are defined by *poles* and *ways*. For example, a simple, one-pole, four-way switch will switch one input signal to one of four outputs. Let's say we had a two-pole, four-way switch. This switch has two sets of independent contacts that can be coupled to one of four positions. Let's say one pole was used to switch a radio output to one of four speakers. To know which speaker was being powered, the second set of contacts could be wired to four LED indicators, marked as 1 to 4. Each LED would then light up, corresponding to it's matching speaker. This setup is shown in Figure 2-11.

Jack Plugs and Sockets

Audio connections are made much neater and easier with the use of miniature 1/8-inch jack plug/jack socket combinations. If you're using a jack plug, you're going to need a jack socket. This size of jack plug is almost always found with the headphones provided for portable radios and cassette players. Now you know the size we're talking about. These aren't the huge jack plugs used with electric guitars. The jack plug has a screw-on barrel, often plastic but sometimes metal. Once you remove the cover, and if it's a mono plug, you'll see two connections. There's a short connection to the center pin and a longer

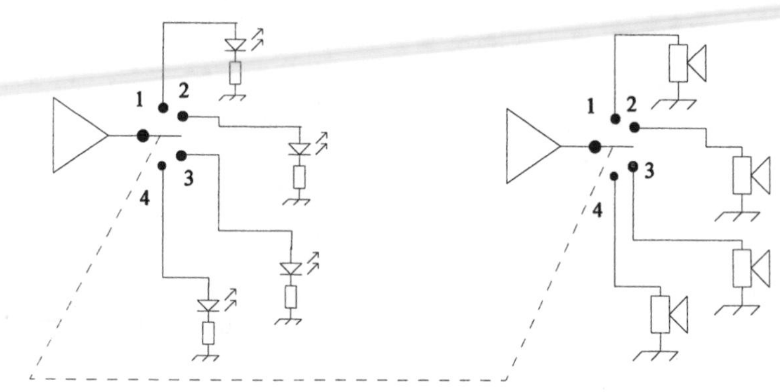

FIGURE 2-11 One-pole, four-way rotary switch

connection that goes to the ground terminal. You can recognize a mono jack plug by the single insulator strip near the end of the jack plug tip. The stereo jack plug has two such insulator strips. Jack sockets come in the normally closed and normally open types. In the normally closed type of jack socket, there are two contacts that are in mechanical and electrical contact, that is, it's normally closed. The action of inserting the jack plug causes the two contacts to mechanically spring apart, so the electrical connection is broken. When you remove the jack plug, the electrical connection is made again. The normally open type of jack socket has two close by terminals that are not electrically connected to each other. When a jack plug is inserted, these two contacts are mechanically brought together and as long as the jack plug remains inserted, the electrical connection is maintained between the two terminals. Figure 2-12 shows the differences for one particular type of popular socket. For the basic application, such as connecting a speaker to a amplifier output, it makes no difference which type is used. But the normally closed type of socket has a spe-

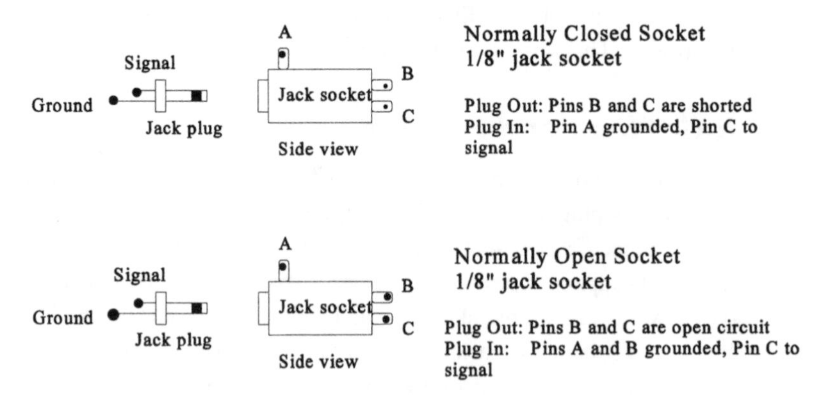

FIGURE 2-12 Jack socket conventions

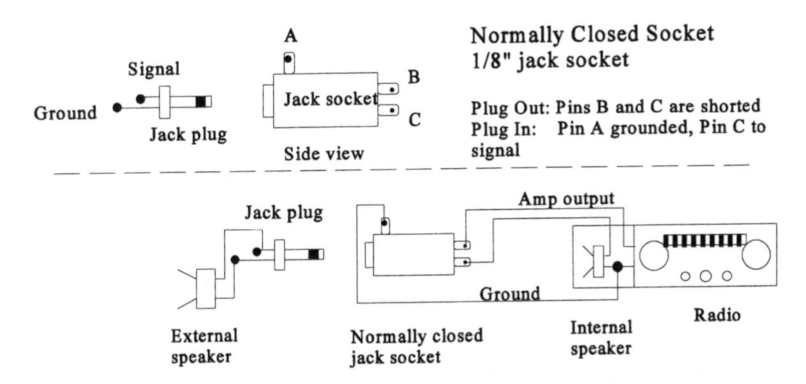

FIGURE 2-13 An example of a normally closed jack socket

cial use; it is used where an amplifier is connected normally to an internal speaker, and when an external speaker is plugged in, the internal speaker is disconnected by the action of this jack socket. Portable radios have the same arrangement, whereby plugging in the external headphones disconnects the internal speaker. This example is seen in Figure 2-13. Like switches, jack plugs and sockets are more complex than they might at first seem.

LEDs

The light emitting diode (LED) is today's solid-state marvel, the equivalent of the filament indicator lamp of years gone by. When I started in hobby electronics, especially in the building of amplifiers, I always had to use filament indicator lamps as power on/off indicators. They took up a lot more space than LEDs, but more critically, the current they drew was enormous. Fortunately, with the advent of the integrated circuit era came also the solid state electronics age, with the LED soon becoming the universal indicator device. Small, light, extremely robust and drawing an economical amount of current, the LED is a natural for panel indicators. In absolute terms, the current drawn is not insignificant, however, but as the rest of the electronics technology speeds ahead to devices that use much less power, the indicator remains locked (at least for the time being) with the LED. Fundamentally, if the LED is to be used as a relatively long range viewing device, current has to be supplied to produce the visible light energy. Typically current through the device is limited with a resistor to just a few milliamps for acceptable viewing. LEDs come in a limited range of colors—red, green, yellow—but by far red is the most common and useful color. They come in different shapes (cylindrical and rectangular) and sizes, from pin-head tiny to jumbo sized, the most commonly used size being something like the size of a TV remote button. There are some special LEDs with very high brightness levels, but they draw more current than the plain vanilla variety, so unless you really need extra high brightness, be careful when you choose your LEDs. The LED package is sometimes marked with the brightness and current values, depending on where you buy your components. Of course, you can always increase the brightness level a good deal by increasing the current up to it's maximum limit, but your battery life will be shortened. There's always a compromise, isn't there? Who's ever heard of a Corvette that's economical to run too.

Integrated Circuit Sockets

Integrated circuit sockets solder into your assembly board and they are a real convenient way of ensuring you never have to go through the painful process of desoldering an IC from a board. You can remove an IC from a board, of course, provided you know what you're doing and have had lots of practice. But why give yourself pain when you don't need to? Integrated circuit sockets come in different sizes—8-pin, 14-pin, 16-pin, and so forth, to match the IC you're using. Make absolutely sure that the socket matches the number of pins on the IC. I've made the error before of installing a 14-pin socket, and to my horror only finding out later, when I'd already built the entire board, that the IC was a 16-pin device. The sizes (14- and 16-pin) are very close. Aargh, the horrors of trying to remove an IC socket!

The mounting convention for IC sockets is important also. The cut-out slot on the socket usually faces to the left hand side, so that pin #1 of the IC is located at the lower left-hand corner. This convention ensures that the IC is plugged in the right way around; otherwise the IC might be destroyed when the power is applied when the device is in the opposite position. While on the topic of the correct handling of IC sockets, I can offer another good tip: solder the pins in the following order. Start with a corner pin and solder it in. Flip the board over and check that the socket is seating flush with the board. If it isn't, you can reheat the pin while pressing down on the socket body. The solder will melt and the force will push the socket down flush. Hold the socket until the solder sets. Then go to the opposite diagonal and do the same. Solder that pin down while pressing the socket down flush. After that the rest of the pins can be soldered in any order. Trying to correct a socket that is not flush with the board can only be done by removing all the solder and starting again. Why should you bother about a flush socket? The appearance of a board that has been put together well is tied with your pride in making something well; a flush socket looks neat.

Assembly Platform

What use is a bunch of components and a neat design if you have nothing to build it on? There are a number of basic options that we'll touch upon.

In electronics hobby magazines you'll often see layout patterns given for producing your own printed circuit boards. I've never liked that idea, getting involved with the hassle of chemicals and so on. Besides, it seems to require an awful lot of effort to build just one board—and one board is all I'd ever want. So what other options do you have? Ready-made prototyping boards in a number of different designs are available from your typical electronics supplier, and they come in a variety of sizes. The pattern on the top surface generally takes the form of a regular array of holes into which components can be inserted and then soldered on the underside. Some prototype boards are also designed to take the pitch of ICs or IC sockets. ICs or IC sockets come in a variety of sizes. The pitch refers to the spacing between the IC terminals. These terminals emerge in a parallel row from either side of the IC. Between the rows of terminals there are, of course, no connections. So a prototyping board that is specially designed to accommodate ICs will have a pattern that matches the underside of the IC, that is, there are no connections running between the IC.

CHAPTER 3

Building a Project Using a Basic LED Circuit

This section is a very easy introduction to describing and building a very simple circuit before moving onto the subject of amplifiers. So by way of an example, the LED circuit is used. What we are going to do is describe a circuit that will turn an LED on and off, very simple I know, but it does make a nice simple start.

Once the schematics for a circuit design have been completed, the exciting project-building process can begin. There is no greater satisfaction to be gained, especially if you're a beginner, than to find that the circuit you've just built actually works. By the same token, you'll be very frustrated if all you've ended up with is a mass of lifeless electronic components.

For the sake of those readers who are new to the construction process, let's run through a test circuit so that we can become familiar with some of the factors that can make all the difference between a working circuit and a dead circuit. At least that way you will know how to avoid some common mistakes, and if your circuit doesn't work you'll have a troubleshooting plan to debug (correct) a dead circuit. We will use the ever popular LED circuit for our tests.

As you can see in Figure 3-1, there are just four components in an LED circuit. They are

- a battery,
- a switch,
- an LED, and
- a resistor.

The circuit function is simple enough; you want the LED to light up when the switch is thrown.

Most errors result from incorrect connections, more specifically errors arising from misconnections. Components rarely if ever are the cause of failures. The greater the complexity of the circuit (the more components there

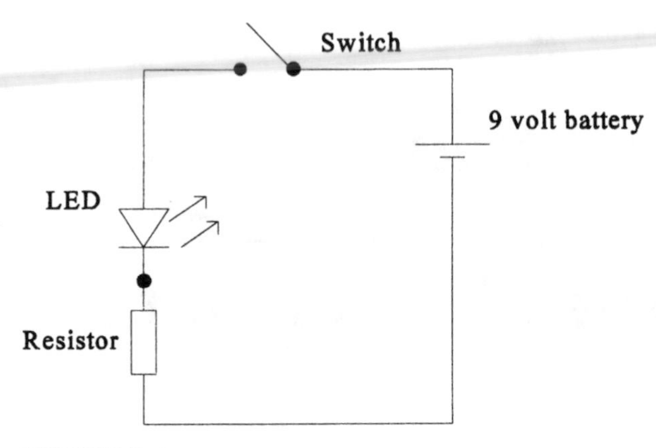

FIGURE 3-1 A basic LED circuit

are), the greater the chances of making an error. That's another reason for going through this exercise with the LED. It's a confidence booster. Start small and build your way up.

The following steps will describe the components we will need to put this circuit together. Each component will be described in turn as we need them, with details of how the connections are made to them, and to work out the correct identity of a component's terminals. The order of the components encountered are:

a. LED
b. 9-volt battery snap
c. resistor
d. 9-volt battery
e. switch
f. circuit board

LEDs

The light emitting diode (usually red) is a really robust device with two solid leads emerging from the body. They're quite close together, so you'll have to bend them out slightly to make a connection. The LED is a polarized device; that is, it can only be connected a certain way to a dc voltage source (a battery). If you look carefully at the body of an LED, you'll notice that one side of the component has a flat edge. The lead closest to that end has to be connected to the negative voltage terminal. In this case the LED is in the forward biased mode, and excess current will flow unless it is limited by a resistor. Typically if you're using a 9-volt battery, a resistor of 4.7 kohms value will give your LED enough brightness yet not cause excess current drain. With battery-operated circuits, current conservation is important if you don't want to be forever changing batteries. Ironically, if you use a resistor with a value that is too low, most of the current drawn from a simple circuit can be from the LED. With the current limiting in place, reversing the LED connection won't cause any damage—the LED just won't light up.

One of the simplest checks can be done within seconds. You need a LED, a resistor (any value between 1 kohm and 10 kohms will do), and a 9-volt battery. Carefully bend the LED leads away from each other. Take one end of the resistor and hold it so that it is touching either one of the LED leads between your thumb and first finger. With a bit of fancy finger wiggling, place the other lead of the resistor in contact with one end of the battery and the other LED lead to the remaining terminal; it doesn't matter which way the connection to the battery is made. The LED will either be lit or not lit. Carefully reverse the connections to the battery. The LED will reverse its state. Not only does this quick test demonstrate the polarized property of the LED; it is also a neat way of identifying how the LED's polarity needs to be dealt with. You can see now that a nonlighting LED could be traced to an incorrect polarity connection.

A 9-Volt Battery Snap

Although it is not shown on the schematic, as is typical, the 9-volt battery snap is an important part of our circuit. A battery snap is the battery connector that clips onto the top of a 9-volt battery. You will typically find the snap nestling inside the back of most small transistor radios operating off 9-volts. What could go wrong with a battery snap? After all, it is just a connector! Well, surprisingly, battery snaps are fairly fragile items (they aren't really components), and that's the problem behind their poor reliability. Underneath the sealed plastic cover above the top of the metal snaps, the two connecting wires are soldered to the metal terminals. These metal terminals are the terminals that mate with the terminals on top of a 9-volt battery. The two connecting wires have a remarkable propensity for breaking off. The only way to find out whether or not your battery snap wires are fragile is to cut away the plastic and investigate. Having seen the wires break off so often, I've come to suspect the snap terminal whenever there's a power loss.

Apart from the weakness of the wires, there's another potential problem with battery snaps. The wires (red and black) leading out from the snap are short and thin (very thin). In most cases they will need to be extended to be of better use. But once you extend the wires, any undue stress on the leads will eventually break the connection between the snap wires and your fragile leads. So it's essential to anchor the leads to the circuit board to prevent undue flexing. Use thin, flexible extension leads (the wires you find in ribbon cable are pretty good). Cover the soldered join with tape so you don't get a short. Strictly speaking, electrical insulation tape is ideal, but since there will hardly be any high voltage or current flowing, any tape will do (masking tape, Scotch® tape, duct tape, whatever's handy). It's a good idea to do a final check once you've finished with the snap's leads. Attach the snap to a fresh 9-volt battery and verify the voltage with a dc voltmeter.

Another thing, the snap contacts can sometimes be too loose, resulting in an intermittently unreliable circuit. Carefully examine the offending terminal and gently bend the hollow-shaped terminal just a shade together to get a tighter grip.

Resistor

Resistors must be one of the simplest, most robust, noncritical components to work with. Resistors are noncritical components because you can often vary their values in a circuit, with the circuit still continuing to function. The recommended value in the parts list can often be varied up or down without detriment to the operation of the circuit. For example, the current limiting resistor used with the LED can be anywhere between 100 ohms and 10 kohms, when a 9-volt supply is used. That's a huge range across which to operate. The leads are robust—almost impossible to damage—and you don't need to worry about polarity dependency. Polarity dependency means that the component has to be connected into the circuit a specific way. A diode is an example of such a component that can only be connected a certain way. Resistors can be connected either way. The value of a resistor is marked on its body; check out the value with an ohmmeter to be sure. If you're using a 4.7-kohm resistor, the markings are yellow, violet, red. The last color band is close to orange, so check it carefully. Yellow, violet, orange will give you 47 kohms, and that certainly won't draw even a glimmer from the LED. So verify your resistor's value with a meter check every time.

9-Volt Battery

I've lost count of the number of times that I have wasted hours of searching for a nonexistent fault only to find that the battery I was using was dead. If a battery is too low, say around 7 volts, discard it. It hasn't got the capability to drive any current. Sooner or later you'll use it in a critical circuit and pick up poor responses.

Switch

As we've seen, the simplest on/off switch has two terminals. But often you'll find other types with four or six terminals turning up. All of these can be used as simple on/off switches, but which terminals are you going to use? Use your trusty ohmmeter again and find a pair of contacts that will show an open circuit/short circuit as you toggle the switch lever. It is a common mistake to use terminals that do nothing at all.

Circuit Board

Typically the circuit board is not shown in a schematic, but like the battery snap, it exists in the final build. Circuit boards in themselves are totally reliable, but problems can arise at the hookup. Shorts can result from any number of problems: solder bridges between tracks (a very common mistake); cold solder joints (just a case of practicing your technique); incorrect solder connections (nothing you can do to prevent that, except check, check, and check against the circuit schematic); and missed solder joints (a common mistake).

So we can see that with just a simple LED lighting circuit, there are a number of areas where things can go wrong. Better to cut your teeth on this check circuit before going on to the projects listed later. For more helpful tips for the beginner, I suggest you check out my first book, *Beginning Electronics Through Projects*. From here we'll be heading into a discussion of ac audio amplifier and related topics.

CHAPTER 4

Audio Pre-amplifiers

Audio pre-amplifiers boost the relatively weak signals that come from tape decks or CD players to a level where they'll be compatible for driving the following power amplifier stage.

Audio signals fall within the so-called audio frequency band, which extends more or less between around 60 Hz and about 20 kHz. This frequency band is generally taken to represent the audio frequency band or the frequency range over which human hearing extends. Younger children and females tend to have a greater ability to hear the higher pitched frequencies by comparison to males and the more senior among us. So a generic audio amplifier would be expected to amplify signals in the range of about 60 Hz to about 20 kHz. I say *expected* because other circuit elements, the speaker, for example, might have a much narrower frequency range, perhaps from a few hundred Hz to just above the level of 10 kHz. That being the case, an audio amplifier need not be designed to have a frequency bandwidth wider than the weakest link in the audio chain. Besides, there is cost and effort involved in striving for wider bandwidths. If you don't have to, don't! For the applications in this book, we are not striving for the hi-fi reaches of audio.

Audio signals generally are emitted by a cassette player, CD player, phonograph or record player, microphone, electric guitar, and so forth. The signals from these devices have something in common: they're low level, a few millivolts or so, and incapable in their raw form of performing any useful function. They certainly do exist, though. Weak signals of the type coming directly from microphones or electric guitar cannot directly drive a power amplifier, so they have no inherent use. Once they are boosted in amplitude by passing through a pre-amplifier, they then are able to perform a useful function. If you attach an oscilloscope to these low level signal sources, you will get a display of the output signal. In order to use those sounds we need to boost the signal up to a more useful level using a pre-amplifier. Even after the signal is amplified, it needs to be fed into a second but different type of amplifier, not a power amplifier. It does provide more gain or amplification; it provides a means of driving loads such as a speaker or headphones. Speakers have a low impedance, so they need a power amplifier to make them function properly. We'll get to power amplifiers later.

With pre-amplifiers we don't need unlimited gain, since too much gain will cause the resulting amplified signal to clip or overload and produce distortion. Typically an output amplified signal level approaching a few hundred millivolts peak to peak is fine. Input signal sources can be considered to have amplitude levels between a few millivolts to perhaps a few tens of millivolts. So the required stage gain between the input and the output would be somewhere between ×10 to ×100. Of course, for very high-level input sources you'll be looking at gains of even less than ×10, because we'll be working with a high level signal to begin. Where the signal is very small, a much higher gain, perhaps ×100 to ×200 will be required, to bring the signal up to the same level of requirement. But inherent noise levels increase with higher gain settings, so that's a good reason to keep the stage gain within control. A figure of ×100 is a really good value to use as a sort of reference value.

The projects outlined in this book use ac signals, meaning signals that undulate and cycle between a positive value and a negative value. If we consider the signal to be a simple sine wave to start with, then we can visualize the signal initially being positive in value on the first half of the cycle and then reversing to being negative on the second half. But the amplifiers under discussion here can be configured either in an inverting mode or a non-inverting mode. Take, for example, a pre-amplifier configured in the inverting mode that is fed with a sine wave input. If you were to assign the input signal to the first channel of an oscilloscope and the output signal to the second channel, you would see the following: When the first half of the input signal is positive, the corresponding output signal is negative. On the remaining half of the signal cycle, when the input goes negative, the output goes positive. A non-inverting pre-amplifier would produce an output signal that is in phase with the input signal. When the input is positive, so too is the output, and the same applies when the input is negative. These signal relationships make no difference to the listener using an ac amplifier, but there is another breed of amplifiers, called dc amplifiers, for which this phase difference serves a purpose. In various electronic project magazines and books you will no doubt see both types of amplifiers mentioned, and for that reason we will cover both types here. Judging from the greater proliferation of inverting mode amplifiers, I'd say they are in the majority.

The Inverting Mode Pre-amplifier

Let's start with a run-through of the more prevalent inverting mode amplifier. We are focusing on using IC op-amps, and the workhorse industry standard is the LM 741. This device has 8 pins or leads commonly in the 8-

pin DIL, or dual-in-line package. It is inexpensive, reliable, and readily available. If your neighborhood electronic parts store stocks only one IC, you can be sure it's the '741. Although this IC is an 8-pin device, our pre-amplifier designs use just five pins: the power and ground connections, the two input pins, and the output pin. Disregarding the power and ground pins, which really exist just to connect power to the device, we're left with three pins that really do the amplifying work. Since these are very important pins, for the sake of future reference, here are the pin designations for the LM 741:

Inverting input	Pin #2
Non-inverting input	Pin #3
Ground	Pin #4
Output	Pin #6
Power	Pin #7

In all our circuit descriptions we'll be dealing over and over again with Pins #2, #3, and #6.

The inverting mode pre-amplifier uses just two components—two resistors to set the gain. That's why the op-amp is so much more versatile as an amplifier than are discrete components. The op-amp's gain is set entirely by the ratio of the two resistors, and its gain is independent of the IC itself. The gain of a discrete amplifier, by comparison, is determined and influenced by a host of circuit parameters, not excluding those of the transistor itself. Thus, designing a stable amplifier is a task unto itself. No wonder the IC revolutionized the design. The op-amp's first resistor, called the *feedback resistor,* is coupled between the output (pin #6) and the inverting input (pin #2). It's termed a feedback resistor because it feeds current from the output back to the input. The second resistor is also attached to the inverting pin, and its other free end goes to the input signal source. Let's call this resistor R1 and the feedback resistor R2. Gain is simply the ratio of the feedback resistor to the input resistor, or R2/R1.

To build a real amplifier we need a few more support components, but we'll come to that soon. For a gain of ×10, R2 could be 100 kohms, and R1 is therefore 10 kohms. We choose 100 kohms for R2, since it is a value that works well. To get a little more gain we could lower R1 to 4.7 kohms for a gain of 21.3, or we could stay with R1 = 10 kohms and raise R2 to 220 kohms for a gain of 22. In practice, R1 can be reduced to a level as low as 2.7 kohms, and R2 can be taken as high as 470 kohms, giving a gain of ×174. Again, you can go higher, but sometimes circuit instabilities may occur. The R1 at 2.7 kohms and the R2 at 470 kohms are nice, stable values for a high gain. As you can see, gain setting is incredibly simple.

The rest of the circuitry needed to get the amplifier to actually function is a little more involved. Let's take a look at the easier part first. Since this is an ac amplifier, the input signal must be capacitively coupled to block dc voltages and pass only ac signals, and likewise for the output. Therefore, we need a series input capacitor coupling the input signal to R1 and an output capacitor extracting the amplified signal from pin #6. Typical values are 0.1 µF for both the input capacitor and the output capacitor.

You hadn't forgotten about pin #3 had you? This is the positive or non-inverting pin. In all ac amplifier designs it has to be taken to what is termed the mid- or half-supply voltage bias point. Since the supply voltage is 9 volts (all the circuits described here use a 9-volt battery for the power source), the held voltage or Vcc/2 bias value is 9/2 = 4.5 volts. To obtain this voltage we simply use (initially) two resistors of equal value. It's the classic potential divider circuit. Again, experience plays a part here in determining what values to choose. Since these two resistors connected in series are going to be connected directly across the power supply voltage, the resistor values should be high, so we don't draw any excess current. Two 100-kohm resistors are a good choice. With a 9-volt supply voltage and a total of 200 kohms connected, the current flowing through the resistors is 9 volts/200 kohms = 0.045 mA (Ohm's Law coming into play), which is really a negligible amount of extra current drain. The midpoint of the two 100-kohm resistors provides the needed 4.5 volts. You can easily check this out by probing the resistor junction and ground with a dc voltmeter. It'll read half the supply voltage. Pin #3 is connected to this junction point. In some instances you will find an additional series resistor buffering the midpoint from pin #3 and providing an extra degree of stabilization.

Finally, since this is an ac amplifier, the mid-voltage point needs to be at an ac ground potential. To do this a capacitor, typically 100 µF, shunts the midpoint to ground. At the ac frequencies under consideration, the reactance of this shunt capacitor is essentially a short circuit. For a capacitance of 100 µF, the reactance at 1 kHz is 1.59 ohms. Elegantly efficient, the simple parallel RC network is a short at ac but a bias provider at dc.

A capacitor has a reactance or impedance, which is a function of frequency. The higher the frequency, the lower the reactance. Reactance is measured in ohms, just like resistance. The reactance value is a function of the capacitance and the frequency that is being considered. When considering audio frequencies, a good choice of a frequency is 1 kHz, we need to select a single frequency to make the calculation for reactance. The capacitor we're using is a large one, that is 100 µF. The reactance works out to 1.59 ohms. This value is so low it is practically a short circuit, and this is, of course, what we need to ensure: that the midpoint of the two bias setting resistors is at a ground potential. The elegantly simple circuit of two resistors and a capacitor therefore, provides us with the required dc bias conditions (Vcc/2) and the required ac bias conditions (ac ground) all in one.

Summary Fact Sheet for the Inverting Amplifier

- The input signal always goes to the negative inverting pin (pin #2).
- The gain-setting resistors always go to the inverting pin, which in this case happens to be the same pin #2.
- The positive non-inverting pin (pin #3) is always fed with a Vcc/2 bias.

Note: When drawing circuit schematics for the pre-amplifier, it's a good idea to follow a convention of having the input signal always go to the "upper half" of the IC. As we know, when the circuit is an inverting pre-amplifier, that pin has to connect to a negative pin. This is a good memory jogger if you're trying to remember which is the negative/positive pin.

The Non-Inverting Mode Pre-amplifier

You will sometimes see this circuit used, but to a lesser extent than the inverting type. Since this circuit is non-inverting, we know that the input pin has to be a positive pin, and following our layout convention, the upper half of the IC has the signal input feed. Note that the pin numbering for the input is upside down as compared with the inverting pre-amplifier.

Let's look at the gain-setting part first. Knowing that in the case of the inverting pre-amplifier there are two resistors for fixing the gain, do we have the same setup for the non-inverting preamplifier? Yes, there is a similar arrangement. There is a feedback resistor, call this R2, connecting the output pin (pin #6) to the inverting pin (pin #2). This is in line with what we've defined previously; the gain-setting resistor goes to the negative pin. But where does the input resistor, R1, go? As we know, the input has to go to the positive pin. Input resistor R1 is taken from pin #2 to ground, via a series capacitor, C1. Gain setting is determined by R1 and R2 and is approximately equal to R2/R1. Typical values for the feedback resistor, R2, would be 100 kohms. Using a value of 10 kohms for R1, we have a gain of ×10. The input feeds into positive pin #3, and because this is an ac amplifier we need the usual input capacitor (C2) and output capacitor (C3). Values typically would be 0.1 μF for both capacitors. But there's an additional section of circuitry to be added. Just as with the inverting pre-amplifier we need a Vcc/2 bias for the positive pin, but where does it fit in this circuit? The half-supply bias voltage goes directly to pin #3. The mid-supply bias voltage is set up exactly as before, using two resistors as a potential divider to split the supply voltage. A stabilizing capacitor connects the potential divider midpoint to ground. A value of 100 µF is commonly suitable. But an additional series resistor is needed to couple pin #3 to the midpoint potential.

The significance and importance of the extra series resistor needed with the split voltage bias supply needs a little introduction. With a casual glance at the circuit it might be missed, but it is very necessary. Without that humble component, the circuit will not function. The incoming signal feeds via the usual series coupling capacitor into the non-inverting positive IC terminal. Here's how to realize the way this part of the circuit functions. Start by tracing the connection path back from pin #3 via the extra series resistor and down to the large electrolytic capacitor to ground. Assume first of all that the series resistor is not there. Capacitors have an ac resistance property called *reactance,* which is measured in ohms and is a function of frequency. For a capacitor value of 0.1 µF and using a typical audio frequency of 1 kHz, the calculated capacitive reactance is given by

$$Xc = 1/(2*\pi*f*C)$$

which works out to be 1.59 kohms. That's a relatively low value, and it drops to an even lower value of 159 ohms at 10 kHz. The input signal coming in through capacitor C2 would be severely shunted by Xc. So the extra series resistor is added. Now the combined effect will be like having a shunt impedance of at least 100 kohms—negligible loading on the input signal. This is a good example of the elegance of electronics circuit design where neat little circuit tricks actually turn an application schematic into a real working circuit. Application notes often provided in the manufacturer's data books, invariably provide bare-bones schematics, merely to illustrate how the circuit functions. These are not real working circuits; often these designs are put together by applications engineers. As a hobbyist, if you attempt to use these application notes as they stand, your success cannot be guaranteed. Unfortunately, some electronics compilations of circuits tend to include these application notes. Perhaps the assumption is that the reader will have sufficient know-how to realize that more components are needed for the circuit to be made to function. Application notes can generally be easily identified because they do not list values for any of the components.

Where the gain is very high, for example, with a gain in the region of ×1000, high-frequency noise or hiss can be reduced by adding a small shunt capacitor across the feedback resistor. A value of from a few 100 pF upward to 1000 pF can be tried. The higher the value of the shunt capacitor, the higher the frequency will be. The noise hiss will be reduced, but so too will the high-frequency brilliance.

Summary Fact Sheet for the Non-Inverting Amplifier

- The input signal always goes to the positive non-inverting pin (pin #3).
- The gain-setting resistors always go to the inverting pin #2.
- The positive non-inverting pin (pin #3) is always fed with a Vcc/2 bias.

Choosing a Safe Range of Component Values

When you need to choose values for either of the pre-amplifiers I've described, you can't go wrong if you follow these rules:

- *Input resistor.* Stay within the range of 1 kohm to 10 kohms; for higher gain you'll want the lower value.
- *Feedback resistor.* Stay within the range of 100 kohms to 1 Mohm; for higher gain you'll want the higher value.
- *Decoupling capacitor across the mid-voltage bias.* Stay with the range of 10 µF to 100 µF.
- *Input capacitors.* In most cases 0.1 µF is the best general-purpose value to use. For a little more bass boost, the input capacitor can be upped to 0.47 µF.
- *Gain-setting capacitor for non-inverting amplifier.* Stay with 0.1 µF, but higher values toward 1 µF will produce bass boost. If a 1-µF electrolytic capacitor is used, the negative end of the capacitor always goes to ground (in the case of a circuit using a positive supply voltage).

Note: Later in Chapter 10, "Construction Projects," you'll see that the two basic amplifier types we've just described are used as the two starter projects. The text in this chapter can be used to supplement the information given in that later chapter. If you want to review amplifier figures now, refer to Figures 10-1 for the inverting amplifier and Figure 10-2 for the non-inverting amplifier.

CHAPTER 5

Audio Power Amplifiers

Power amplifiers take an incoming signal and provide it with enough drive to run a low impedance speaker.

In this chapter we're not looking at those massive megawatt amplifiers so beloved of heavy metal rock guitarists or the power-hungry audiophile. No, this is the hobbyist's introduction to audio power on a much smaller scale; in fact, the power output is no more than a watt. But 1 watt is more than sufficient power to produce output sound levels that are adequate for the hobbyist who wants to experiment with amplifiers. For normal listening levels we rarely exceed a watt of power. The focus here is on the ease with which power amplifiers can be built, using no more than a handful of components.

Way back in time, before the advent of the integrated circuit—in fact, before even the introduction of transistors or printed circuit boards—amplifiers using vacuum tubes were constructed by a point-to-point wiring method. Heavy output transformers, the warm glowing ranks of monolith vacuum tubes on a metal chassis, topped many hi-fi enthusiast's tabletop in the 1950s. As transistors became available, initially the germanium types (now defunct) became popular amplifier (power and pre-amp) building blocks, offering a much more compact unit, but, more important, they were battery operated and, hence, portable.

Transistor-based power amplifers could be constructed using a significant number of parts, often requiring to be matched, and requiring critical bias setup conditions. And an output transformer was always needed for the push-pull class B type of output stage that was in vogue then. A simpler class A power amplifier could be constructed, requiring less parts, but the current drain was horrendous because the class A stage drew current regardless of whether or not there was an input signal present. At least the class B stage had a reasonably low quiescent current consumption, turning on only when there was an input signal present.

A class A amplifier is one that is operating in a mode such that the output current flows during the full cycle of the input signal. This is the simplest amplifier configuration, but it is inefficient as the output current flows even in the absence of an input signal, hence, the current drain from the power supply is high. A class B amplifier is one that is operating in a mode such that the output current flows only during one half of the input signal. This is a highly efficient mode of operation as the output current is directly proportional to the input signal and is almost zero for zero input signal.

By the early to mid-1960s, silicon transistors started to replace the germanium originals, but this development didn't simplify power amplifier designs. Not until the arrival of integrated circuits and in particular the ultimate low-power, low-cost, easy-to-use audio power IC—the National Semiconductor LM 386—did the power amplifier really take off. Today the majority of lower-power audio circuits contain the LM 386. This IC is ridiculously simple to use and it has revolutionized the entire process of constructing a simple power amplifier. It can almost be considered a "drop-in amplifer on a chip," considering the small number of parts needed to get the power amplifier up and running. The benefits of the LM 386 can only be fully appreciated if you've been through the pain of trying to procure, let alone build, and set up a pre-386 power amplifier.

Why do we need a power amplifier in the first place? All audio devices (cassette players, radios, stereos, CD players, and so forth) ultimately feed into a speaker as a load.

"Speakers" and "loads" tend to be used synonymously in the context of power amplifiers. A speaker is basically many turns of fine wire wound into an arrangement called a voice coil. When the coil is energized by feed with the signal from a power amplifier's output, the electrical signal is transformed into mechanical energy such that the speaker's cone (this is the characteristic cone shape of a speaker) vibrates in sync with the electrical signal. In doing so, sound waves are produced. This is the sound we hear coming from a speaker. If you were to apply a small dc voltage (a volt or so) to a speaker, you can see the speaker cone move out with one particular polarity and in when the polarity is reversed. This illustrates the conversion of electrical to mechanical energy. Because of the fact that the electrical portion of the construction of the speaker consists of a coil of wire and the electrical resistance of that wire is low, the speaker is thus referred to as having a low impedance.

When any signal is shunted by another component (this could be the resistor in the simplest case), we describe that signal as being loaded down or just loaded. The component that does the "loading" is obviously referred to as the "load." The lower the impedance of the load, the greater the shunting effect on the signal, that is, the more the signal will be reduced (this is undesirable). Load, in a more general sense also refers to any entity that a signal feeds into. Consider the load as the final destination of the signal—it has to end up somewhere, that is the load.

A speaker, from an impedance point of view, is a low-resistance load, typically around 8 ohms. In order to drive a speaker, you need a current source that has the drive capability for the low impedance speaker. That's where the power amplifier comes in. It's more than just having a high-voltage source. The pre-amplifier, for example, generates adequately high voltage, but it is incapable of driving a low-impedance load. What would happen if you were to connect a speaker directly across a pre-amplifier? The output voltage would effectively drop to an ineffectual level. The power amplifier provides the needed current. The power has to come from somewhere, and that's how the power supply, or battery in this case, serves its purpose: the higher the output or sound level, the more current is drawn and, correspondingly, the shorter the battery life. Distortion will set in when the battery voltage drops to a level that's below the mimimum operating voltage.

Prior to the advent of the LM 386, building a power amplifier, even a modest low-power unit, was a daunting task, especially when the aim was just to have a capability to drive a speaker and to sustain a reasonably long battery life. I remember my own attempts to build power amplifier projects; they were bulky, difficult to construct, and overall unsatisfactory. Using two external components, coupling capacitors at the input and output, the LM 386 can be immediately up and running, a feat that no other device to date (as far as I know) can better. Current consumption with a 9-volt battery is gloriously low, around 6 mA.

Gain settings are easily done with the aid of a resistor-capacitor combination between pins #1 and #8. The capacitor is generally fixed at a fairly large value of 10 μF. The higher the series resistor, the lower will be the gain. At the limit, when the resistor is removed, the gain mazimizes to around ×200. The data in Table 5-1 provides resistor values for various gain settings.

TABLE 5-1 LM 386 Gain-Setting Resistor Values. Rx = 390 Ohms for Lowest Hiss (Gain~65)

Input frequency = 1 kHz sine wave, 10 mv peak to peak. Rx = gain setting resistor in series with 10 µF capacitor, connected between pins 1 and 8.

Input	*Output*	*Gain*	*Rx*
10 mv	200 mv	×20	10 kohms or oc
10 mv	300 mv	×30	2 kohms
10 mv	400 mv	×40	1 kohm
10 mv	500 mv	×50	632 ohms
10 mv	600 mv	×60	442 ohms
10 mv	700 mv	×70	319 ohms
10 mv	800 mv	×80	252 ohms
10 mv	900 mv	×90	183 ohms
10 mv	1000 mv	×100	147 ohms
10 mv	1100 mv	×110	105 ohms
10 mv	1200 mv	×120	75 ohms
10 mv	1300 mv	×130	53 ohms
10 mv	1400 mv	×140	37 ohms
10 mv	1500 mv	×150	18 ohms or sc

Circuit details on building an LM 386 power amplifier are given later in Chapter 10, "Construction Projects."

CHAPTER 6

Simple Filter Designs

Understanding Filters

Filters are just circuits that allow you to control the frequency response of, for example, an amplifier. Two basic types of filters are covered here: the low-pass filter, which passes low frequencies and gives you more bass sound; and the high-pass filter, which passes or allows the high frequencies to be transferred unimpeded to the next stage, and gives you more treble sound.

Once you've read through this book, you'll probably notice that the most common circuit configuration for an op-amp-based amplifier is invariably the inverting mode circuit, where the gain-setting components consist of a simple feedback resistor and a input resistor. In the majority of cases, that's about all we're concerned with. But if you want to look further, especially at other publications, you'll come across circuit techniques for manipulating frequency response via filter networks. Some of these are simple and some complex. We'll take a look here at the addition of a very simple component that will work wonders with the frequency response of an amplifier. The more practiced technician will probably scoff at the simplistic design. But for me if it works, then that's good enough, especially for beginners. Many of the more complex arrangements are horrendously difficult to understand. The treatment here is meant to teach you from the ground up, and once you get the grasp of these filter designs, you'll soon be able to recognize other circuits that act as filters.

In a filter context we are talking about the frequency response of a circuit to an ac signal. The response of the circuit, especially in the case of an op-amp-based design, can be transferred to the response of just one component. We can't be referring to a resistor, because resistors are not frequency sensitive. So it has to be the capacitor, the only other logical device. The property we're looking for in the capacitor is its reactance, that is, its ability to have a resistance that varies as a function of frequency. The reactance is given by the equation

$$Xc = 1/(2*\pi*f*C) \text{ ohms}$$
where $\pi = 3.142\ldots$
f = frequency in Hertz
c = capacitance in Farads

Using this relationship, the reactance can be calculated for any combination of capacitance and frequency. What do you notice about this equation? The reactance is inversely proportional to the frequency; that is, as the frequency increases, the reactance drops. This is the basis of the filtering property.

Because the calculation process is a little tedious, Table 6-1 provides a matrix of reactance values that have already been calculated for you. Five very common capacitor values are used to set up the table. Across the audio frequency band, we have six spot frequencies spanning the low end (100 Hz) to the upper end (20 kHz). Look at a capacitor with a value of 0.1 µF (which is used in most of the circuits in this book as a coupling device either for the input signal or the output signal). If we take 1 kHz to be a good representative single frequency for the audio band, then we see the reactance is 1.59 kohms. What does this mean? Well, 1.59 kohms is quite a low value. It's a reactance in series with the input signal or the output signal, so there's no significant attenuation of the signal (which is what we want). Too high a value would give us a reduced signal. What value is too high? A value somewhere in the region of 10 kohms and above could be considered to be a value at which we will begin to see a reduced signal. Now let's look at the reactance for a frequency of 100 Hz. This time the reactance is much higher (actually by a factor of 10), 15.9 kohms, and this value will affect the signal. If the signal has 100 Hz components, then using a capacitor of 0.1 µF will eliminate some of these frequencies. The higher frequencies, above 1 kHz, are easily passed since the reactance reduces appreciably to the sub-kohm region. To all intents and purposes, any reactance value below a kohm could be considered to be a short circuit.

TABLE 6-1 Capacitive reactances for common capacitor values across the audio frequency band

Xc	*f = 100 Hz*	*f = 1 kHz*	*f = 5 kHz*	*f = 10 kHz*	*f = 15 kHz*	*f = 20 kHz*
C = 100 pF	15.9 Mohms	1.59 Mohms	318.27 kohms	159.13 kohms	106.09 kohms	79.57 kohms
C = 1000 pF	1.59 Mohms	159.13 kohms	31.83 kohms	15.91 kohms	10.61 kohms	7.96 kohms
C = 0.01 µF	159.13 kohms	15.91 kohms	3.18 kohms	1.59 kohms	1.06 kohms	795.67 ohms
C = 0.1 µF	15.91 kohms	1.59 kohms	318.27 ohms	159.13 kohms	106.09 ohms	79.57 ohms
C = 1 µF	1.59 kohms	159.13 ohms	31.83 ohms	15.91 ohms	10.61 ohms	7.96 ohms

What happens if the capacitance is changed? As expected, there is a reactance shift with a change in capacitance. Going to a higher capacitor, say 1 μF, now produces a reactance of 159.13 ohms at a frequency of 1 kHz frequency—lower than the previous case. So more of the signal will be passed. In reality, for most of the frequencies in the middle part of the audio band, the difference won't be noticeable, at least to the human ear.

But let's look at the lower frequency of 100 Hz. This time the reactance with the 1 μF capacitor is only 1.59 kohms. This level of reactance is sufficiently low to pass more low-frequency components than if we were using the 0.1 μF capacitor. If you were running an audio setup that had the capability to reproduce the low 100 Hz frequencies, then the increase in bass response would be very noticeable. Going to even higher values (10 μF, 100 μF) produces even more bass. You'll notice later in the power amplifiers projects that the output coupling capacitor is always large (100 μF)—that's so there will be no loss of the more hi-fi bass frequencies. Using a smaller capacitor would (apart from generally reducing the signal level) make the sound more tinny, and you'd lose the lower bass notes. Of course, this discussion applies mainly to music reproduction. With speech, producing more bass might contribute to a more boomy sound, in which case you might want to deliberately cut the bass frequencies instead.

Low-Pass Filter

Now that we've covered the response of the capacitor to frequency changes, let's go back to the basic amplifier circuit (Chapter 5). Figure 6-1 shows a representative low-pass filter design. There's a capacitor, C1, straddling the feedback resistor, R2. That's the only addition to the basic circuit. The plot at the side shows how the gain falls off as the frequency is

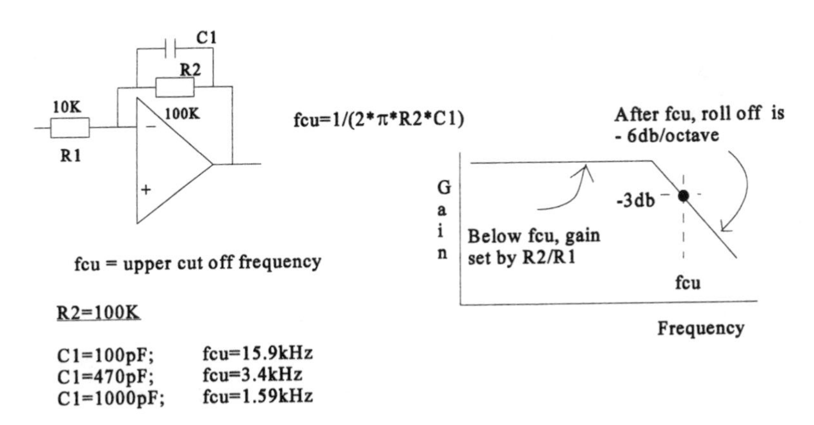

FIGURE 6-1 Low-pass filter

increased. The point at which the gain falls by 3 decibels (db) is called the *upper cutoff frequency point*. A decibel (or db) is just another convenient way of representing the ratio of two signal levels. When we are considering a voltage ratio, a value of 3 db, or actually –3 db to be more accurate, means that the ratio between the second signal to the first signal is 0.7. If for example, the second signal had a magnitude of 7, and the first signal had a magnitude of 10, then the ratio would be 7/10 = 0.7. In this case we can consider the second signal to be the "falling high frequency," and the first signal to be the "flat mid-band gain" level. The absolute signal levels don't matter at all when you're working with decibels. So we could assign a value to the mid-band gain, and for convenience we could call it ×10. This would be the gain value set up by the ratio of R2 to R1. So if R2 were 100 kohms and R1 were 10 kohms, this would give us the desired gain of ×10. Now, if we were measuring the output signal on an oscilloscope and increasing the frequency as we went along, at some point the gain would start to fall off (decrease). It would become less than ×10. The point where it falls to ×.707 (that is, 7.07) is defined as the upper cutoff frequency. The ×.707 (or –3db) point is a convenient marker point to establish fcu, the upper cutoff frequency. Without predefining a marker point, we couldn't locate fcu. If you look carefully at Figure 6-1 you'll notice that the frequency response is labelled as rolling off at –6 db per octave. We know that –6 db is some reference to the ratio of two signals. Actually –6b is a voltage ratio of 0.5. The octave descriptor refers to a frequency that is twice the original frequency we've starting with. Let's say the upper cutoff frequency is 10 kHz. Then the frequency that is an octave above it is just 2 × 10 kHz or 20 kHz. So that fixes the two frequency points. Now we'll go back to the –6 db. As we've said, that's a ratio of 0.5. So if the gain at fcu were say 8 (any number will do), then the gain at an octave higher would be half that, or 4. Knowing the two frequency points (10 kHz and 20 kHz) and the two gain values (hypothetically 8 and 4) we can draw a straight line between the two coordinates. That's how the plot is generated. A more detailed diagram is given in Figure 6-2.

In addition to the octave, which was defined above as the ratio between two frequencies such that one was twice the value of the other, there is another ratio used. This is called the decade. Whereas the octave refers to a frequency that is twice the original frequency, the decade refers to a frequency that is ten times the original frequency. If our reference frequency were 10 kHz, then a frequency a decade higher would be 100 kHz (somewhat outside of the audio frequency band). The decibel falloff is different, as we would expect. For the same response slope, we can therefore describe it as having a fall off of –6 db/octave or –20 db/decade—the two are equivalent. Both descriptors mean the same thing. You'll see both units of measure

db	Gain
-3db	0.7
-6db	0.5
-10db	0.3
-20db	0.1

Octave = original frequency x 2

Decade = original frequency x 10

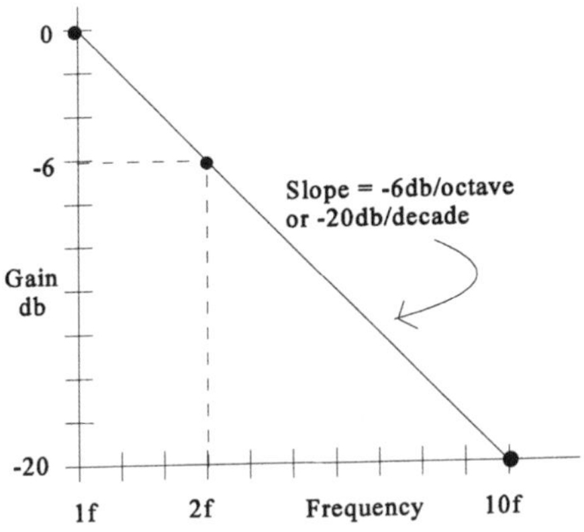

FIGURE 6-2 dbpix

used interchangeably. The fcu point is also shifted by the actual capacitor value. Higher-value capacitors will lower the point at which fcu occurs, or, put another way, they will provide more high-frequency attenuation. A low-pass filter thus can be seen as a design in which the high-frequency response is tailored by the choice of suitable feedback components (R2 and C1).

Finally, we come to Table 6-2, which shows the parallel effect of having the capacitor, C1, straddle the feedback resistor, R2. The capacitor by itself already has a falling reactance as the frequency is increased, and this factor together with the use of input resistor R1 means the gain will fall off with increasing frequency. That is why this circuit is called a low-pass filter; the low frequencies are passed unattenuated; whereas the higher frequencies are progressively more and more reduced. The effect of the fixed resistor, R2, which is needed to set up the correct dc operating conditions, is merely to contribute to the overall impedance. We are effectively looking at a parallel combination of two resistors, and that's how the values in the table are calculated. When any two resistors are combined in parallel, the total effect is always governed by the lower of the two values. You can see this relation in the matrix of values plotted (Table 6-2).

High-Pass Filter

At first glance the high-pass circuit doesn't seem obvious, and it isn't. The setup is different from that of the low-pass filter, still simple but different. As the starting point, we'll look at the basic non-inverting amplifier. For simplicity, we will ignore the dc gain-setting components since they're not needed for the analysis. So without any gain components, the amplifier simplifies to a buffer, which is what we see in Figure 6-3. First of all, as we expect for an ac amplifier, there's a capacitor feeding in the signal. But since

TABLE 6-2 Parallel capacitance/resistance impedance calculations

Capacitive reactances						
Xc	*f = 100 Hz*	*f = 1 kHz*	*f = 5 kHz*	*f = 10 kHz*	*f = 15 kHz*	*f = 20 kHz*
C = 100 pF	15.91 Mohms	1.59 Mohms	318.27 kohms	159.13 ohms	106.09 kohms	79.57 kohms
C = 1000 pF	1.59 Mohms	159.13 kohms	31.83 kohms	15.91 ohms	10.61 kohms	7.96 ohms
C = 0.01 μF	159.13 kohms	15.91 kohms	3.18 kohms	1.59 ohms	1.06 kohms	795.67 ohms
C = 0.1 μF	15.91koh ms	1.59 kohms	318.27 ohms	159.13 ohms	106.09 ohms	79.57 ohms
C = 1 μF	1.59 kohms	159.13 ohms	31.83 ohms	15.91 ohms	10.61 ohms	7.96 ohms

RC impedances						
Xc	*f = 100 Hz*	*f = 1 kHz*	*f = 5 kHz*	*f = 10 kHz*	*f = 15 kHz*	*f = 20 kHz*
100k//C = 100 pF	99.375 kohms	94.082 kohms	76.092 kohms	61.409 kohms	51.477 kohms	44.311 kohms
100k//C = 1000 pF	94.082 kohms	61.409 kohms	24.144 kohms	13.726 kohms	9.592 kohms	7.373 kohms
100k//C = 0.01 μF	61.409 kohms	13.726 kohms	3.083 kohms	1.565 kohms	1.048 kohms	789.389 ohms
100k//C = 0.1 μF	13.726 kohms	1.565 kohms	317.260 ohms	158.877 ohms	105.978 ohms	79.507 ohms
100k//C = 1 μF	1.565 kohms	158.817 ohms	31.820 ohms	15.908 ohms	10.609 ohms	7.960 ohms

this is a high-pass filter design, there's an extra component located at the input—the resistor R1, which is connected between the positive IC pin and ground. The resistor R1 basically provides a shunt path for the signal that is being applied to the op-amp. As before (in the low-pass filter example), there's also a frequency roll-off plot. This time the area of activity centers on the low-frequency point. Using the same definitions as before, the lower cut-off frequency, fcl, is indicated on the plot. The formula for calculating fcu is as before, except that the values for R1 and C1 refer to the components at the input to the amplifier. Since the numbers will be exactly the same as in the previous case, we can use the data from Tables 6-1 and 6-2.

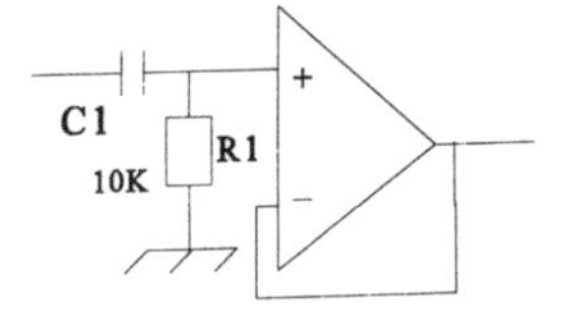

fcl=lower cut off frequency

R1= 10K

C1=0.1uF; fcl=159 Hz
C1=0.47uF; fcl=34 Hz
C1=1uF; fcl=15.9 Hz

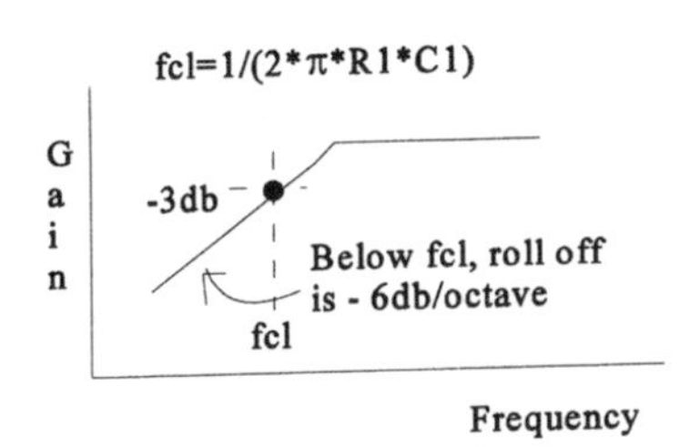

FIGURE 6-3 High-pass filter

We need to decide on a suitable value for R1. This is quite a critical value when we are configuring the high-pass filter. If the value for R1 is too low, the signal coming from C1 will be heavily attenuated, and little will pass through the system. But if the value of R1 is too high, then negligible filtering will take place. So a good compromise value (gleaned from experience) is 10 kohms.

Having fixed R1, we can see the effect that varying C1 will have on the calculation of fcl. Our standard value for C1 is usually 0.1 µF, so using this value as a starting point, fcl works out to be 159 Hz, with R1 = 10 kohm. That's quite a good low-frequency cutoff point, so we should get a decent sort of a signal (that is, some bass should come through). But we still need to find out what happens if we change the value of the capacitor. Upping the value of the capacitor lowers the fcl, which translates into boosting the bass frequencies. If we use the value C1 = 0.47 µF for C1, the fcl drops to 34 Hz. This is good. Even more bass can be gained by going lower to C1 = 1 µF, and fcl then sinks to 15.9 Hz, definitely bass heavy.

Buffer Low-Pass Filter

Although I have already covered the low-pass filter, it's interesting to note that the simple action of swapping R1 and C1 turns the buffer high-pass filter into a buffer low-pass filter, as is shown in Figure 6-4. It's an easy trick, but some thought is needed in choosing the components. Now, the signal enters through resistor R1, and the shunt component is C1. We know that the value of R1 can't be too high, since a high value results in too much attenuation and the loss of signal. An R1 with too low a value approaches a short circuit and deviates from the low-pass filter's ability. So a good value to use for R1 is 1 kohm. Once we've determined the R1 value we can plug in values for C1 and calculate fcu.

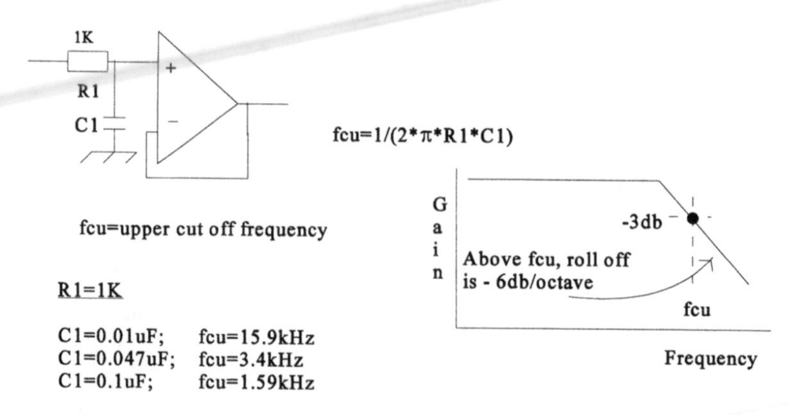

FIGURE 6-4 Low-pass filter

The C1 capacitor also has limitations. If you choose a value that is too high you will get too much shunting signal loss. Typically, a value of 0.1 µF would be about the highest you should use. Let's start with a smaller value of 0.01 µF and see what happens. With this value, fcu works out to be 15.9 kHz, a good value, giving lots of high-frequency response, that is, not much high-frequency attenuation or cut. Going higher to 0.047 µF causes the fcu top to drop to 3.4 kHz. That's fairly low, and the effect can be heard. Finally, down at the 0.1 µF value fcu is now way down at 1.59 kHz. The high frequencies are definitely gone. Simple tone controls often use a shunt capacitor like this with an added potentiometer in series to provide the ability to make adjustments.

CHAPTER 7

Circuit Schematics

Electronics schematics are the equivalent of the architectural plans used for building your home: they tell you what it is you're building, what gets connected to what, and how it all comes together. A schematic is the accurate representation on paper of what is physically connected in the real project. Since the ability to design your own circuits is tied into the ability to use schematics, a starter lesson in the makeup of schematics is a good idea.

Take the simplest case of a flashlight circuit. The schematic is shown in Figure 7-1. There are three components in a regular flashlight: a battery (or batteries), a lightbulb, and a switch. Inside the flashlight the batteries are connected in series, and one end is attached to one of the bulb terminals. The other

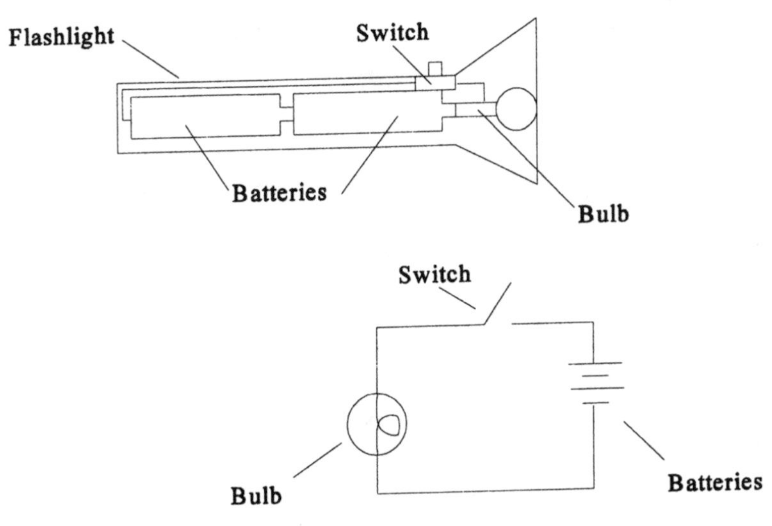

FIGURE 7-1 Flashlight schematic

end of the battery stack is connected to a switch. The other end of the switch goes to the second bulb terminal. When the switch is closed, the light comes on. In the schematic we see symbols for the battery, the switch, and the bulb, together with the interconnecting wiring. By following the schematic we are able to make the necessary connections to duplicate the flashlight circuit. That's basically the reasoning behind schematics. They use a set of standard symbols to represent the physical components and to show how they are interconnected to create the physical circuit. With the simple flashlight example we're starting with something we know (flashlight circuitry) and working from there to draw the schematic. You'll use a schematic to build each project in this book. In all the schematics we follow the same conventions, so as you progress through the book, you'll gain familiarity with schematics.

All the supply voltages used have a positive polarity. This is represented by a horizontal line laid out across the top of the page. This line represents the positive battery terminal. The ground line, or negative voltage lead, is represented by a second horizontal line at the bottom of the page. All the rest of the circuitry is drawn between these lines. Circuits always need a switch. It is inserted in the positive supply line and located to the far right of the supply line. The only connection to the far right of the switch is the supply voltage. Everything else appears to the left of the switch.

The next most important thing we need to know is when the power is on; otherwise we're liable to run down the battery. Fortunately, the LED (light emitting diode) is a superb, small, robust, solid state lamp. Before the advent of the LED, indicator lamps had to be of the filament variety, consuming large amounts of current and restricting their use to line power driven circuits. All LEDs require a current-limiting resistor for correct operation, but apart from that, you need only wire the LED across the power supply, on the "down side" of the power switch (the terminal that is not connected to the power supply [the battery in this case]).

Although there is not a critical need to have smoothing capacitors across the power line for the simple circuits shown using just one IC (since the supply source is a battery), using smoothing capacitors is good practice for cases where the dc power is supplied from a voltage adapter. Additionally, more complex circuits often required the use of smoothing capacitors across the supply line. For that reason—that is, to establish good design practices—two capacitors are included, a large value electrolytic and a small disc ceramic capacitor across the power supply, connected in parallel with the LED. Let's recap what we know so far. These components associated with power supply are common to all the circuits shown in this book. Let's assume you want to construct a circuit you've designed yourself. Make a start with the components I've described in this chapter and that'll cover your power supply needs.

The next schematic stage concentrates on the actual device of interest itself. For the purposes of this book, we're looking at the integrated circuit. The device we build will define the IC device type we need to use. Once that is chosen, there are the rest of the components, some specific to achieving that function and the rest necessary for bringing the IC to the correct operating condition. For example, an IC for an audio pre-amplifier application running off a single battery (9 volt) will always require a Vcc/2 bias connected to the positive input terminal, regardless of the specific gain required. In the more common inverting mode, two resistors determine the gain. And as far as audio applications are concerned, there always has to be a capacitor in the input and at the output.

So here's how to check out a new circuit and see if it's an amplifier: There will be an input signal marked on the schematic, perhaps from a microphone, radio source, cassette deck, and so forth. If there is a capacitor in the input line, it's an ac circuit. And there's a capacitor feeding the output signal too. If the input goes to the inverting pin (if the IC is an LM 741, this is pin #2), then this is an inverting circuit. There should be a resistor linking the output terminal to the inverting pin, and a second resistor (generally of smaller value) also connected to the same inverting pin and feeding off the input capacitor. The ratio of the feedback resistor to the input resistor determines the gain. For example, the commonly used values of 100 kohms for the feedback resistor and 10 kohms for the input resistor produce a gain of ×10. As we now know, this is an inverting ac amplifier running off a single supply. Thus, there has to be a half bias voltage supplied to the non-inverting IC terminal. This is usually a network of two equal resistors connected across the supply with the midpoint taken to the non-inverting terminal. Generally there's an electrolytic capacitor also shunting the split-supply voltage.

Let's summarize this last analysis to see how easy it is to analyze what might at first appear to be a new and therefore strange circuit and to logically break it down into its different functions. We are looking at the IC itself, since we've already covered the basic power supply components.

1. The IC symbol is a triangle on its side, pointing to the right.
2. There are two inputs on the left and one output on the right of the triangle.
3. There's a negative inverting terminal and a positive non-inverting terminal.
4. There's a feedback resistor connected from output to inverting terminal and an input resistor attached to the inverting terminal, so this is an inverting amplifier.
5. Knowing it's an inverting amplifier, there is then a split supply feeding the non-inverting terminal, using two resistors and a capacitor.

Four of the most popular ICs used in projects in Chapter 10 are shown in Figures 7-2 through 7-5.

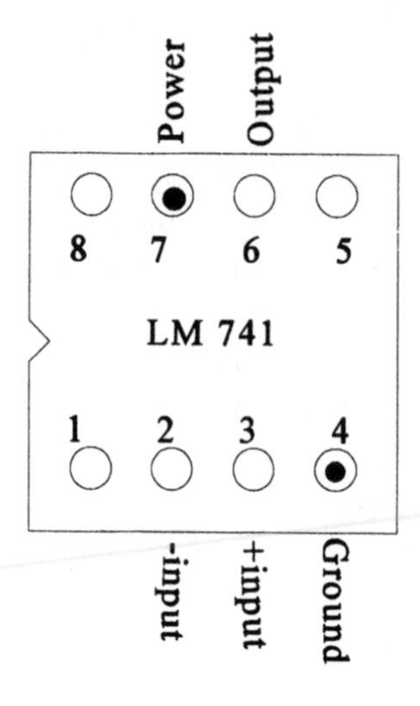

FIGURE 7-2 LM 741 pinout. The LM 741, an 8-pin device, is by far the most widely known general-purpose IC. Shown here are the pins that are critical for the projects featured in this book.

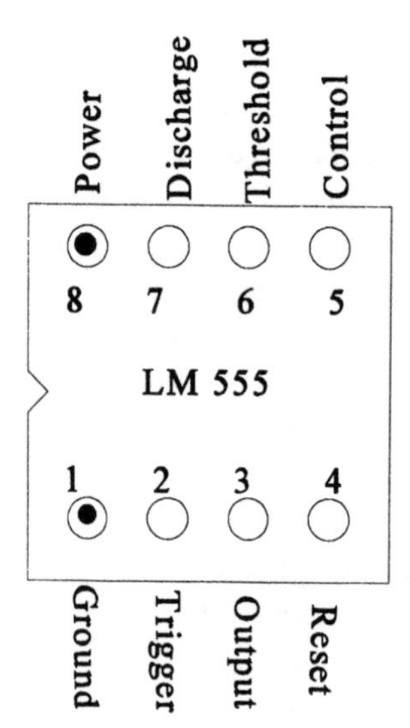

FIGURE 7-3 LM 555 pinout. The LM 555 is an incredibly versatile 8-pin IC, which makes the design of signal generator circuits as easy as ABC.

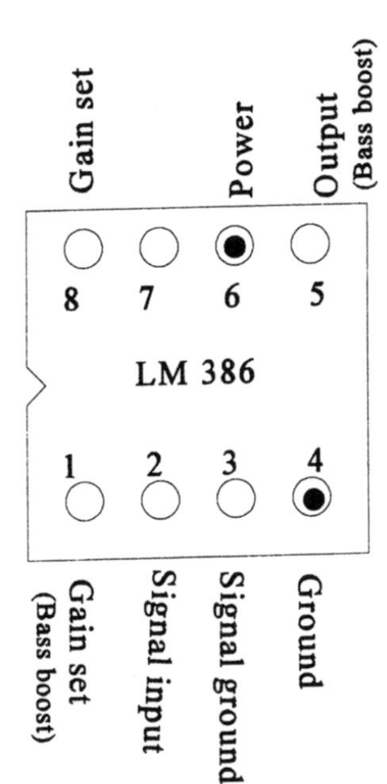

FIGURE 7-4 LM 386 pinout. The LM 386 is without parallel—the ultimate simple-to-use power amplifier. Packing a more than generous amount of drive in a simple 8-pin package, it is superb.

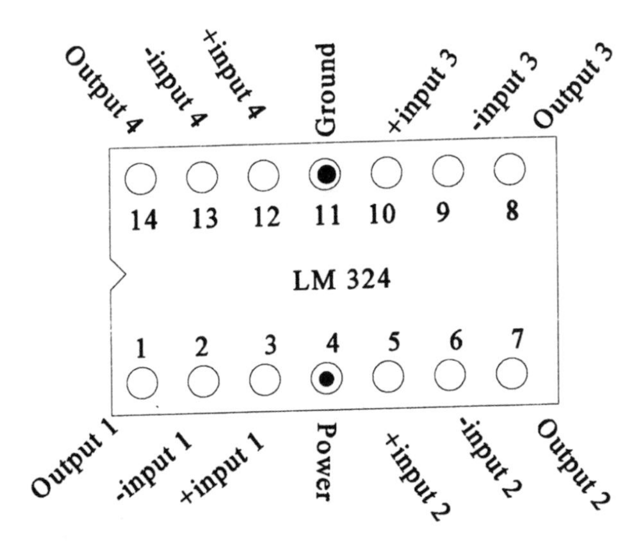

FIGURE 7-5 LM 324 pinout. The LM 324 is an unusual device, a quad op-amp, in a larger 14-pin package. It is often used in circuits for which several op-amps are required, since this one IC takes up less space than four separate units.

CHAPTER 8

Test Instruments

You can dig deep into your wallet for a bevy of the latest exotic test instruments, but they'll certainly come with a hefty price tag. For the beginner to electronics, and especially hobby electronics, where the focus is really on getting enjoyment from building circuit projects, there are really only four basic instruments that you need in order to do fundamental testing—and we're talking about testing for the sake of getting a circuit to function or troubleshooting a circuit that is not working, rather than taking a lot of detailed measurements.

Testing involves either making measurements at various points in the circuit or injecting a signal at the input stage and seeing what comes out at the output stage. If a circuit is not working, more often than not the problem is only a matter of a simple error involving a wrong connection. Having test instruments at your disposal will make the task of troubleshooting so much easier for you. Being familiar with how to use these instruments is not a bad idea, and it makes your understanding of circuits so much better and, I hope, your work more interesting.

The four test instruments I'd like you to get to know are the multimeter (very essential), the signal generator (expensive, but covered to an extent by one of the circuits described in Project 9 of Chapter 10), the function generator (nice to have but also expensive), and finally the oscilloscope (really nice to have, expensive, but there's no cheap alternative). We'll discuss these four instruments one at a time so you can see what each one does and whether or not you really need it. Let's make a start with the simplest and cheapest.

Multimeter

The multimeter measures three basic electrical elements in both the dc and ac modes: voltage, current, and resistance in both the dc and ac modes. But as far as we're concerned, that is, with regard to the circuit projects described in this book, only the dc ranges are of relevance, so we'll concentrate more on those. Multimeters are very affordable, and prices for the lower performance beginner units are not expensive. One of the first and major characteristics of multimeters is that they are available in two types—analog and digital. Which should you buy? The older style analog multimeter is very easily recognized by its needle pointer, which swings across a graduated face on the instrument. The analog device is very much a moving mechanical instrument. The newer, but not necessarily better, digital counterpart has no mechanical moving parts, and the display is purely numerical, just like your digital watch. Going back to the analog instrument, in response to an applied signal, such as dc voltage (from a battery), the pointer needle will swing clockwise and display the input voltage, which is measured across a calibrated scale. A large central dial indicates the voltage range. Typically you might find the following dc voltage ranges available on the meter: 2, 20, and 200 volts. So let's say we're measuring a 9-volt battery. The correct setting to select is the one that is greater than the voltage you're trying to measure. This is really important if you're using an analog meter. If you select a lower setting you can do serious damage to the needle, because in taking the reading the needle is likely to hit the end stop with significant force. It's a good practice therefore to start with the highest setting (say 200 volts) and work down. That way you'll avoid causing expensive damage.

Another very important consideraton is to make sure the polarity of the test leads is correct, that is, that you connect the right lead to the right terminal. The test leads are color coded red and black. Red is by convention associated with the "hot" or positive voltage, and black with the negative or ground voltage. The terminals on the multimeter itself are also marked (colored) to indicate which leads should be connected where on the meter. Basically you match the red lead to the red terminal and the black lead to the black terminal. Having done that, you can then go on to do some dc voltage measurements.

So let's say you want to measuure the state of a 9-volt battery. Once you've connected the red lead to the red terminal, momentarily touch the black test lead to the black terminal on the battery. I always take this safety precaution because if by any chance I've made a connection mistake somewhere, the meter needle will kick in the wrong (counterclockwise) direction, and I will immediately know that I need to withdraw the test probe from the battery terminal.

Multimeters have a large function selector switch located on the front of the instrument. You turn it to the appropriate item you want to measure, typically this would be voltage, current, or resistance.

Resistance measurements can be tackled by turning the center selector switch to the resistance range. Always disconnect both test leads from the item you're measuring before changing to a different type of measurement, again to prevent damage. For resistance measurements you might have settings such as 2, 20, 200 kohms. You may select any resistance setting when you're measuring resistors—there won't be any damage done by selecting an incorrect setting. First of all you need to zero the resistance setting. Do this by shorting the leads together with the meter switched to any resistance range. The meter needle will swing across to the right. You can then use the small zero adjust potentiometer on the multimeter to adjust the needle so that it lines up with the zero mark on the resistance scale. The small zero adjust potentiometer is used to zero the needle on an analog multimeter, when it is set to measure resistance and the test leads are shorted together. You do this procedure each time before you take a measurement on a resistor in order to get a correct resistance reading. Each different resistance range needs to be zero-adjusted before it is used.

Select a resistor that you want to measure. Read off the color code so you've got an idea of what the reading should be. Place the test probes across the resistor, and the needle should swing across to the value corresponding to the resistor's coded value. Make sure you are not using your fingers to hold the probes to the resistor, because your body's resistance will act in parallel with the resistor value and produce a false reading. If the resistor's value is outside the setting you're using, the needle will not deflect sufficiently for a reading to be made. There is no potential of doing damage. The meter leads can be swapped when you are making resistance measurements. Measuring resistance with a meter is especially useful when the colors on a resistor are difficult to distinguish and you need to be sure about the resistor's resistance value.

Measuring current is trickier because the circuit you're building needs to be configured a little first. You need to set up a small test circuit to see how current measurements are made. Connect up a 9-volt battery, an LED, and a current-limiting resistor of, say, 10 kohms. The LED will be fairly dim, but you're using it for demonstration purposes only. We know from Ohm's Law that the current flowing through the LED will be the voltage (9 volts) divided by the resistance (10 kohms). This works out to just under an mA of current (that is, 9/10,000 = 0.9 mA). So set the multimeter to the current setting that's higher than an mA. If you're not sure, just set the multimeter to the highest current setting. You will very likely need to place the red test probe in a different socket. The instruction book for your particular meter will tell you how to proceed, or else there will be markings on the multimeter to show which terminal socket to use.

Going back to the LED circuit, remove the positive voltage connection from the LED and touch the current meter's red probe to the positive battery supply. Momentarily touch the black test probe to the free LED end. All being well, the meter needle will deflect to about an mA, to show the current drawn by the LED. If the needle swings off the scale or in the opposite (counterclockwise) direction, quickly break the connection before any damage is done to the meter needle. The current meter is usually protected by a fuse, so if you find the current range is totally dead, check out the fuse (situated inside the meter). More than likely the fuse will be open circuit. Not too sure how to check out the fuse? Easy. Remove the fuse, set the multimeter to the ohms range, and check the continuity between the fuse terminals. A good fuse will give a low resistance value; if the fuse is blown there will be no reading. Especially with the type of fuses where the fuse wire can't be seen, a continuity check is the only way to determine whether or not the fuse is viable. Incidentally, a continuity checker is merely a simple tester for verifying the electrical connection between two points. That's exactly what an ohmmeter does, so if you've got a multimeter, you don't need to buy a continuity checker, too. When a multimeter is set to the resistance range (it doesn't matter which one), the test probes can be used as a continuity tester. If you apply the probes between two points of a circuit, you know you have a connection if the meter reads zero ohms (that's the same as a short circuit). The simple LED circuit we've been discussing can also be used as a continuity checker. Make a break anywhere in the circuit and bring out two flexible leads that are connected to the break you've just made. When you connect them to the connection you're trying to verify, such as the fuse, the LED will light up if the connection is sound, and vice versa.

Apart from a few minor differences, the digital multimeter is very similar to its analog counterpart. The digital instrument will tolerate voltage overloads, and in the presence of excess voltage, the display will indicate that you have a signal overload. Reversals in polarity will show up as a negative sign before a reading is displayed, so you will know right away that the voltage has been reversed, accidentally or otherwise. In either case, there will be no damage done to the digital instrument.

When making resistance measurements with a digital multimeter, there is no need to perform the zeroing operation, as there is with an analog device. And if a resistance signal is out of range, just switch to the next higher range; that is, if you're trying to measure a 10-kohm resistor on a 2-kohm maximum range, the display won't be able to display a reading, so you will have to switch to the higher range to get a reading. In the case of taking current measurements, the digital instrument is also tolerant of polarity reversals, but a current overload can blow the multimeter's fuse, so take care and determine beforehand whether or not your meter can handle the expected current.

With all the benefits of the digital unit over the analog instrument, there's a very good reason that the analog instrument should be considered. With analog instruments, their pointer needles make it very easy to read changing voltage values. The eye is totally suited to following the movement of the pointer needle and instantly gauging whether the voltage is increasing or decreasing. In addition, rates of change of voltage can also be very easily determined. The digital instrument is great for displaying very precise information, but it's practically impossible to look at a slew of rapidly changing digits and determine the extent to which the reading, say voltage, is going up or down or to get an idea of the rate of change of the voltage level. The analog unit makes possible excellent visual tracking of rapidly varying voltage levels, and that is why you still see the banks of analog meters in recording studios.

Having made these positive observations about analog meters, though, if I had to choose just one instrument, then the digital multimeter has more benefits overall. The unit tends to be robust—most likely it will stand up to being dropped accidentally; whereas the analog meter would very likely cease to function if it makes contact with the floor. There are other exotic functions that more expensive multimeters also perform, but voltage, resistance, and current measurements are the only ones you really need.

Signal Generator

The signal generator produces a clean sine-wave signal, typically in the audio frequency range, and it is used to test the purity of audio amplifiers. When you amplify a sine wave, the sound it produces is like the clear tone of a bell, a sound rarely encountered in today's overdriven heavy metal rock music world. But for test purposes the audio amplifier we are testing should amplify the incoming signal and not produce any distortion, so to start with we need the clean test signal that can be provided by the signal generator's sine wave.

A sine wave is considered a "clean" signal, unlike a square wave, which is "loaded with harmonics" and produces a very distorted sound. A 1 kHz sine wave produces a 1 kHz signal. A 1 kHz square wave will produce the 1 kHz fundamental wave as well as many other higher-order harmonics. The second harmonic, for example, would be 2 kHz, the third harmonic, 3 kHz, and so on. Generally, the odd-ordered harmonics (3, 5, 7, and so on) go to make up the square wave. The signal generator has the ability to vary the signal frequency, from a few tens of Hz to 20 kHz or so if the signal generator is tailored for audio frequencies, and it can also vary continuous amplitude from a few millivolts to perhaps a few volts.

If you've got a new amplifier you want to check out, set the signal generator to a suitable audio frequency—1 kHz is a good value—and turn the signal amplitude way down low in the millivolt region. Switch on your amplifer and feed in the test signal. With the volume control turned up appropriately (a comfortable listening level) you should get a clean signal from the speaker, since the input signal has increased in amplitude by passing through the amplifier. At some point when the input signal is too large for the amplifier to handle, the amplifier will start to distort, and at that point you will know approximately how much gain the amplifier can produce without distorting.

The feature set offered by a commercial signal generator would be nice to have, but if you just need to produce a test tone, and if budget is a concern, a signal generator is way too expensive to buy. If you want to build your own tone generator, there's a nice circuit project included in Chapter 10 that produces an acceptable compromise to a pure tone—and at a fraction of the cost of buying a signal generator. The tone generator you can build also has frequency and amplitude controls for maximum flexibility. The circuit is a simple one-IC circuit based on the popular LM 555 timer device. This tone generator is an ideal test signal generator for evaluating any audio device.

Function Generator

A function generator produces three distinct signals: a square wave, a triangular wave, and a sine wave. The square wave is a repetitive pulse train that flips between a zero value and a positive maximum. The square wave has its minimum value sitting on the zero voltage level and its maximum value going in the positive direction. Typically, a square wave would lie between 0 and +5 volts. The +5 volt value would be the positive maximum value. A triangular wave looks like a sine wave except the voltage rises and falls in a linear fashion; the peak voltages are not rounded off as they are with a sine wave. The sine wave is the purest signal; it follows the mathematical sine wave function. The amplitude and frequency controls on a function generator provide the usual flexibility needed to check out various electronic systems. The square wave signal is very useful for checking out digital circuits especially when the frequency is slowed down so that the transitions can be tracked by eye. The triangular wave, also called a *ramp wave,* has a somewhat more limited use, especially when it comes to audio device testing. The sine wave, as we've mentioned before, has many audio applications because of it's spectral purity.

As far as audio amplifiers are concerned, the function generator has the edge over the signal generator because the function generator can also produce the sine wave signal that the signal generator produces. The absolute spectral purity of the sine wave from a signal generator is not really needed for the simple audio amplifier circuits contained in this book. So long as the frequency range being covered by the sine wave is in the audio frequency band, and it is, then that's really all we want.

The function generator has the additional feature of automatically sweeping the frequency across a band that you select. Apart from a nice ability to produce a cacophony of weird spacelike sounds, I can't say I've found any benefit in this feature.

Although we don't feature any digital circuits in this volume, the function generator's digital capability in the square wave mode is really useful because it can be used to test deficiencies in an amplifier. Being able to slow down the pulse repetition rate to a level at which the eye can resolve it allows you to "see" the pulse transitions taking place in logic switching circuits. Logic switching circuits typically change or switch in value between 0 volts and +5 volts. Used in conjunction with an oscilloscope, an amplifier you are testing is fed a square wave, and any high-frequency or low-frequency deficiencies in your amplifier will show up as clear deviations of the output pulse shape from an ideal square wave. Since the square wave is really composed of a fundamental frequency and many higher-order harmonics (integer multiples of the fundamental frequency), you can get a quantitative view of the amplifier's high-frequency capability by looking at the rising and falling edges of the output pulse. The more limited the amplifier's high-frequency capability is, the more the edge of the pulse will be rounded off, instead of being square like the ideal square wave. But this is really only a visual indication. A much more elegant method involves using sine-wave input and an oscilloscope and measuring the gain of the amplifier while increasing the input frequency. The point at which the output signal starts to fall off can be very clearly seen.

By using the signal generator with its variable frequency capability and an oscilloscope as the test instruments, we can determine the maximum frequency response of an amplifier quite easily. To get a quick visual indication of when the high frequency response of the amplifier starts to fall off, you can watch both the input signal from the signal generator and the output signal from the amplifier at the same time on a dual channel oscilloscope. As you increase the frequency of the input signal, there will come a point at which the output signal from the amplifier will start to fall or decrease. This is because we have started to approach the high frequency limit of the amplifier. This information tells us how good the amplifier is in reproducing high frequency signals.

Oscilloscope

The oscilloscope is an extremely useful display instrument that produces an image of any repetitive signal, such as the sine wave and square waves we have discussed in this chapter. Controls on the front panel of the instrument allow you to adjust the height and the width of the signal. Dual channels are standard features, so both the input and the output signals of an amplifer can be displayed simultaneously. Not only can the amplifer gain be measured—by comparing the relative amplitude of the output signal to that of the input signal—but distortion can be seen as soon as it crops up. As you increase the input signal amplitude, you can see the point at which the output signal starts to distort or clip. It is well worth spending the money to get an oscilloscope so that you can have this kind of versatility in testing.

The oscilloscope's operating principle is based on a glowing spot that's produced on the face of the tube, almost identical to your regular TV screen or computer monitor. An internal sweep generator in the oscilloscope, called a *time base,* drags the glowing spot on the face of the cathode ray tube across to the right, and this action is repeated many times, over and over again. The drag or sweep rate can be adjusted using external controls on the oscilloscope. At high enough sweep rates, because of the persistence of vision of the eye (the eye retains an image of an object for a short time after the object has disappeared), it appears that a continuous straight line is being drawn on the screen. Variable time base settings allow for all manner of signal frequencies to be displayed on the oscilloscope, up to the maximum frequency response of the instrument.

Generally, budget-priced oscilloscopes start with frequency responses around the 10 MHz region, which is more than adequate for measuring audio signals. The signal you are testing is fed into one of two amplifiers, called *Y-amplifiers,* since this is analogous to the Y-axis on a graph. Front panel controls for adjusting the gain of the internal Y-amplifier allow you to adjust the incoming signal to fit the face of the screen. If a sine-wave signal were being displayed without the internal sweep time base switched in, all you'd see would be a straight vertical line, which is not much use at all. But when the sweep generator drags the signal across the screen, you're effectively seeing the signal, the sine wave in this case, sort of stretched open; the signal is being accurately displayed. When the internal sweep time base is locked to the frequency of the incoming signal, the signal appears to be stationary on the screen, so measurements such as amplitude and frequency can be made. You can measure the amplitude of the signal by counting the number of vertical screen graduations the signal covers and multiplying this by the Y-amplifier scale factor. For example, if the scale factor or gain factor is 1 v/cm (volts per centimeter) and the signal is 3 cm high, then the peak-to-peak signal amplitude is $3 \times 1 = 3$ volts.

To calculate the frequency, first you measure the distance between two repeating points on the sine wave, say the distance between positive peaks. Let's say this is 10 cm, and the time base is set to 1 millisecond (ms)/cm. The total period is then $10 \times 1 = 10$ ms. By taking the inverse of this value, that is,

1/10 ms, we get the equivalent frequency of 100 Hz. That's how the oscilloscope is able to give you a measure of the frequency of the incoming signal. Of course, the signal has to be repetitive so that you can measure two consecutive points on the waveform. Pulses can also be measured to determine their frequency. The pulse duty cycle, which is the ratio of the on period to the total on and off period can be easily measured off the screen. There is no other way to measure signal characteristics that is as simple as using the oscilloscope, so I definitely consider this device to be a very versatile test instrument. Since it is costly, before investing in one you'll need to gauge how much use you're likely to get out of it. I would put it as number 2 on my list of buys after the basic multimeter.

Summary

1. Start with a good multimeter, digital if you prefer, since they're more modern, more robust, and probably there's a wider selection to choose from.
2. It's always good to have a second meter, and for greater flexibility an analog multimeter's a very good complement to your digital mainstay.
3. You can always make do with a home-brew signal source, but a oscilloscope needs to be a bought item. So put it up there on your wish list.

CHAPTER **9**

Audio System Hookups

You've built a collection of test projects, and maybe you're wondering how you can hook them all up together. This chapter provides a nice summary—call it a reference guide with tips and cautions—if you want to put all your projects together. I provide step-by-step instructions so you can't go wrong. Remember, always connect up all the items first before switching on the power. Then switch on the power in sequence starting with the item closest to the speaker first; that is, switch on the power amplifier first, then the pre-amplifier, then the signal source. To switch off the power, go in the reverse direction. Switch off the source first, then the pre-amplifier, and finally the power amplifier. What you want to avoid is having a signal source connected to a unit when the power to that unit is switched off.

Testing a Speaker

Requirements: Signal generator, power amplifier, speaker
Connection mode: Signal generator to power amplifier; power amplifier to speaker

There is some value to testing whether or not a speaker has been damaged, perhaps because too much power has been applied to it, or maybe it has been physically abused. If you're getting distortion in an audio hookup, it's really important to know where the distortion is coming from, and that means that you need to make sure that all the component parts are sound. Even when all the components in a hookup chain are good, you can still get distortion by overdriving some of the stages. We'll start with the speakers.

The speaker has two terminals, and these have to be connected to the output terminals of a power amplifier. The terminals can be connected either way round; it makes no difference. What you will usually see in a commercial speaker that is mounted in an enclosure are polarity markings on the back of the speaker, usually black and red. These markers are meant to help you get the phase right with stereo amplifiers; the red speaker terminal goes to the red amplifier output terminal, and the black speaker terminal goes to the black amplifier terminal. (Since we're dealing with only a single amplifier in this project, this phasing requirement doesn't come into play.) Speakers also have a power rating and impedance value (typically 8 ohms). But for the low power levels we're dealing with here it doesn't matter what the speaker rating is. You will always get a better sound reproduction if the speaker you are using is properly mounted in an enclosure, instead of just "bare."

To test a speaker:

1. Connect up the speaker to the amplifier output terminals.
2. Connect the test signal generator source to the amplifier input, but do not switch on the power. Set the amplitude control to the low position and the frequency control, if there is one, to around 1 kHz.
3. Turn the power amplifier volume control to the low position. Do not switch on the power yet. You will have to match the output connector from the signal generator to the type of input connector you have on the power amplifier.
4. All being well, switch on the power amplifier first. You may hear a click from the speaker.
5. Then turn on the signal source. Depending on the relative gain control settings and the gain of the amplifier, you should get a sound from the speaker. If you're using a sine wave or near equivalent as a signal source, the sound output should be clean and distortion free.

Testing a Power Amplifier with a Piano-key Style Tape Player

Requirements: Tape player, power amplifier, speaker
Connection mode: Tape player to power amplifier; power amplifier to speaker

The piano-key tape player is a mono machine used predominantly for recording speech rather than for high-fidelity reproduction (this task is better handled by the Walkman type of tape player). The piano-key-style tape player leaves much to be desired in terms of producing a high-fidelity signal;

it is practically devoid of any bass response, but it is clear enough for speech. There's nothing wrong with that, since voice recording is what the tape recorder was designed for in the first place. But the machine is very robust mechanically and much more suited to heavy usage than the Walkman. The output signal comes from the earpiece socket, which is generally a 1/8-inch miniature jack socket. So an easy connection cable to use for this test is a short length of cable, twisted twin wires or screened cable, with a 1/8-inch miniature jack plug attached at either end—assuming that you have the same size jack socket on your power amplifier (usually the case). If you want to use a socket/plug combination to connect the speaker input to the power amplifier, using this jack plug is a good idea because this plug provides a much more reliable connection than you would get by playing around with loose wires. The tape player has a high output signal level, which is controlled by the volume control dial, so start with the volume setting low.

To test the piano-key-style tape player:

1. Connect the speaker to the power amplifier output.
2. Connect the tape player output to the power amplifier input.
3. Keep the power amplifier volume setting low.
4. Keep the tape player volume setting low.
5. Turn on the power amplifier first. You might hear a click from the speaker when you switch on the power amplifier.
6. Turn on the tape player. You should get a signal from the power amplifier speaker. Adjust the power amplifier for more volume. If the signal level from the tape player is too high, there may be some distortion.

Testing a Power Amplifier with a Walkman-style Tape Player

Requirements: Tape player, power amplifier, speaker
Connection mode: Tape player to power amplifier; power amplifier to speaker

The Walkman-type tape player is a high-fidelity unit, and the headphone output socket is a stereo 1/8-inch miniature jack socket. You cannot substitute a mono jack plug in the stereo socket. Assuming that your power amplifier has a mono 1/8-inch jack socket, you'll need to make up a special cable that has a mono jack plug at one end and a stereo jack plug at the other end. You can buy a regular cord with the same type of jack plugs at each end, cut one off, and replace it with the correct type. Or you can start from scratch and solder a mono jack plug on one end and a stereo jack plug on the other end of a length of screened cable. Mark these accordingly, otherwise you're likely to mix them up. The mono jack plug has a single-signal tip, whereas the stereo jack plug has a double-signal tip (ignoring for the time being the ground connection). The power amplifier output can be connected to the speaker with a mono jack socket/jack plug combination.

To test the power amplifier using a Walkman-style tape player:

1. Connect the speaker to the power amplifier output.
2. Connect the tape player output to the power amplifier input.
3. Keep the power amplifier volume setting low.
4. Keep the tape player volume setting low.
5. Turn on the power amplifier first. You may hear a click from the speaker when you switch on the power amplifier.
6. Turn on the tape player. You should get a signal from the power amplifier speaker. Adjust the power amplifier for more volume. If the signal level from the tape player is set too high, there may be some distortion.

Testing a Power Amplifier with a Record Player

Requirements: Record player, power amplifier, speaker
Connection mode: Record player to power amplifier; power amplifier to speaker

Although you don't see a lot of record players or turntables these days, since they've been somewhat overtaken by CD players, there are still some die-hards hanging on to their vinyl collections. If you've got a record player, here's how to connect it up and test it out.

We'll start with a brief explanation of how record players work. A fine stylus attached to a cartridge makes contact with the record grooves. As the record rotates, the stylus moves vertically in sync with the grooves; the stylus, in turn, pushes against, typically, a special piezoelectric ceramic cartridge. Every time this type of cartridge is squeezed, a small electrical signal is produced. This minute signal is what is amplified. There are variations on the ceramic cartridge, such as the moving-coil cartridge, but the principle is essentially the same; a vertical movement produces a signal output. Ideally because of the low signal input it's best to make the connection between the record player and the power amplifier with screened cable. The hum pickup will be worse if you use regular wire. The connection to the record player is done underneath the turntable platter. Trace back the color coded wires from the cartridge head and solder the appropriate wires to the pickup terminals. Take note which of the connections is grounded. Check that the stylus is actually there and not damaged, otherwise have it replaced.

To test a record player with a power amplifier:

1. Connect the speaker to the power amplifier output.
2. Connect the record player pickup output to the power amplifier input.
3. Keep the power amplifier volume setting low.
4. Turn on the power amplifier. You might hear a click from the speaker when you switch on the power amplifier. Gently tap the head of the cartridge shell. You should detect a clicking in the speaker output; if you do, that's a good sign.

Testing a Pre-amplifier

Requirements: Low-level signal source, pre-amplifier, power amplifier, speaker

Connection mode: Low-level signal source to pre-amplifier; pre-amplifier to power amplifier; power amplifier to speaker

The pre-amplifier has a lot of gain; it is supposed to create gain after all, so a low-level signal source is needed to test out the pre-amplifier. A microphone is a good low-level signal source for testing a pre-amplifier (as is a record player). The inexpensive type of microphone supplied with a piano-key tape player is ideal. It should be the type of microphone that doesn't require a voltage bias (as the electret microphone does); otherwise there will not be a signal forthcoming.

Trial and error is the only way to find out what kind of microphone you have, unless you have an oscilloscope at your disposal. To test a microphone using an oscilloscope, couple the microphone's leads to the oscilloscope's input, turn up the gain, and tap gently on the face of the microphone. You should get some indication of a signal output if the microphone is a basic carbon or crystal type of microphone. One of the circuits included in this book, the "Project 12: Microphone Test Set" in Chapter 10, can be used to distinguish the two different types of microphones, so if you've already built that project, use that circuit to identify your microphone.

To test a pre-amplifier:

1. Connect the speaker to the power amplifier. Turn the volume control down low.
2. Connect the pre-amplifier to the power amplifier, but don't switch on anything at this stage. Turn the volume control down low.
3. Connect the signal source into the pre-amplifier. If the signal source is a microphone, you can use a simple connection of a 1/8-inch miniature jack plug; a jack plug might already be on the microphone's screened cable. If there isn't a jack plug on the screened cable, you can easily solder one on, making sure that the ground screened lead attaches to the larger ground connection terminal on the jack plug. I'm assuming that your pre-amplifier has a 1/8-inch jack socket at the input. If it's got something different, then mate it up with the appropriate plug. Personally I like to stick with the miniature jack plug/socket combination because it's very easy to use, takes up little space, and is a standard fixture on a lot of commercial audio equipment.
4. Turn on the power amplifier.
5. Turn on the pre-amplifier. You should get a definite sound from the speaker. Adjust the volume controls to bring up the sound level. If the setting is too high, there will be positive feedback howl from the speaker, especially if the microphone is brought physically close to the speaker.

The Potentiometer

Audio circuits, whatever their function, invariably include a potentiometer for basic volume control purposes. Therefore, I dedicate a little extra space here to explaining the inner electrical workings of the potentiometer. Details of a very simple experimental hookup are shown in Figure 9-1, using just a 9-volt battery, a 10-kohm potentiometer, and a multimeter. The input voltage, which is close enough to 9 volts, depending on the state of the battery used, is fed to the two outer terminals as shown. The output is always taken from the center terminal (that's easy to remember). The remaining connection to the multimeter can be made at either of the two outer terminals. Since the circuits we are dealing with usually run off a positive supply voltage, the common or ground terminal is the negative voltage. As the potentiometer is rotated clockwise, the wiper terminal voltage will increment smoothly.

What is happening inside the potentiometer as the control knob is rotated? The ratio of the resistances existing between, say, the upper terminal and the wiper and the lower terminal and the wiper is being continually var-

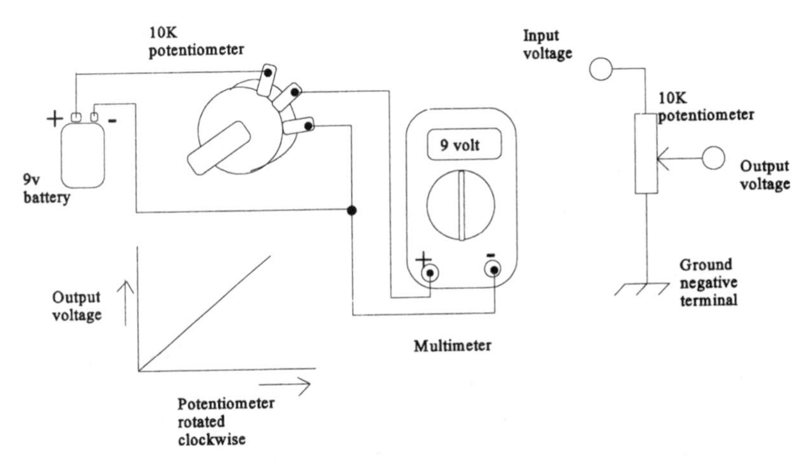

FIGURE 9-1 Basic potentiometer function

ied as the shaft is rotated. These ratios listed in Table 9-1, based on two resistors—the upper resistor R1 and the lower resistor R2—show what the potentiometer is. R1 is connected to R2, and the junction is the point at which the output voltage is taken. The input voltage is applied between the free end of R1 and the free end of R2.

The gain, which is the ratio of the output voltage to input voltage, has a maximum value of 1; all other values are less than 1. The resistor values clearly show (Table 9-1) what the ratios should be to produce the desired attenuation figure. In the simplest case, if you wanted to reduce the input voltage by half, then the values of the two resistors should be equal. All other

TABLE 9-1 Potential divider resistor values. Signal applied to R1. Output taken from junction of R1, R2. R2 grounded.

R1	*R2*	*Gain*
0	10 kohms	x1
1 kohm	9 kohms	0.9
2 kohms	8 kohms	0.8
3 kohms	7 kohms	0.7
4 kohms	6 kohms	0.6
5 kohms	5 kohms	0.5
6 kohms	4 kohms	0.4
7 kohms	3 kohms	0.3
8 kohms	2 kohms	0.2
9 kohms	1 kohm	0.1
10 kohms	0	0

values can be read off from the table. If you've got a particularly large signal, say 1 volt, and you need to reduce this by a factor of 10 before you feed the signal into an amplifier, then a value of 9 kohms for R1 and 1 kohm for R2 will produce the desired effect. How are the ratios worked out? It's quite simple. The output voltage is determined by the ratio of the lower resistor (the one that's connected to ground) to the sum of the two resistors. So in the example shown in Table 9-1, the sum of the two resistors is always 10 kohms to keep things nice and simple for the arithmetic. So working down the list, the first row's gain (×1 gain) is obtained by dividing R2, which is 10 kohms, by R1 + R2, which, again, is 10 kohms. So 10/10 = 1. The next entry down, ×.9 gain, is yielded by R2 , valued at 9 kohms, divided by R1 + R2, which as always is 10 kohms. Incidentally, as we know, the sum of R1 + R2 = 10 kohms (just for this example), then R1 is easily calculated as the difference of R2's value from 10 kohms. You can go down the list of the other gain values and verify for yourself that the ratio of the resistors is correct, always adding up to 10 kohms. The sum could be set at any number. If you used 100 kohms (another popular potentiometer value), then the values for R1 and R2 would be calculated the same way. The general formula is

$$\text{Vout} = \text{Vin} \times (\text{R2}/(\text{R1}+\text{R2}))$$

Given Vin, R1 + R2, and knowing what you want Vout to be, you can calculate the value for R2. Once you've got that, R1 is easily determined.

CHAPTER 10

Construction Projects

Perhaps in the past you've dabbled with building an audio amplifier from discrete components, you know, a bunch of transistors, capacitors, and resistors. Or maybe you've purchased a kit of parts and a ready-made printed circuit board and built a project. Or perhaps you're like me and have put in many years in the electronics business and cut your teeth on vacuum tube amplifier kits.

The integrated circuit has been around since the early 1960s, when Jack Kilby first brought the integrated circuit to life at Bell Laboratories. Remember building your first model airplane: accurately cutting balsa fuselage sections, gluing fragile sections together, wondering if the tissue was going to tear, and then finally the greatest expectation—would it fly? I went through the model airplane phase; most of what I made didn't fly too well, or for too long, for that matter. And then later I saw ready-to-fly airplane kits; snap together wings and tail to the fuselage and engine (rubber or gas powered) and off it flew—and much better than I could have done.

The integrated circuit is a very close analogy to those airplane kits. You can spend an inordinate amount of time and effort building an audio amplifier from discrete components, and it may or may not work as planned (or expected). But put together an audio amplifier based on an integrated circuit and I guarantee it'll work. Not only that, it'll probably work better than its counterpart made up of discrete parts. In today's world of the ubiquitous IC, this is the way to go for amplifier ease of construction.

One of the marvellous aspects of using IC-based designs is that the amplifier's performance is only affected by essentially two external components, and then only by the ratio of these components, and not by their actual values. Discrete amplifiers are affected by the values of the actual components used and also by the transistors themselves, but this is not so with the IC amplifier. You couldn't build an amplifier using just one transistor that could match an amplifier built using an IC-based amplifier, which internally consists of many transistors. The advent of the integrated circuit has taken us a long way toward making amplifier construction much more simple and reliable. These comments are restricted to the scope of this book, where very simple amplifier designs are considered.

Before we go to the first project, let's take a look at specific portions of the circuit schematic, which in some cases, will repeat all the way till the last project. All the circuits are powered by a single 9-volt battery with the positive terminal supplying the power rail. Conventionally the power rail is drawn at the top of the schematic and the applied voltage is drawn on the right-hand side. You can go to the schematic, look for the 9-volt symbol and satisfy yourself it's in the correct location. Next in from the battery is the switch. The switch is always the very first component in from the supply voltage. To the left of the switch is the LED and associated resistor combination. The resistor limits the current that goes through the LED, and the value of the current is a compromise between saving current and being able to see the LED. A value around 4.7 kohms is a good starter value. Feel free to vary it, or if you've got another value that is close to this, use that instead. The LED is used as a power "on" status indicator, and that's a very useful function because without some indicator the battery could rapidly drain if the power were left on inadvertently. You'll probably come back to your workbench the next day and find the battery totally drained. Even at a few mA of current a 9-volt battery doesn't last that long. Two capacitors decouple the power supply to ground. These are located close to the LED. Physically they could be quite a distance away, but it's the electrical connection that counts. These two components enhance the stability of the supply line when a transformer-derived dc voltage is used, for example, that from a line voltage adapter. It's a good design practice to include them, and that's why they're here. The values are commonly 0.1 μF and 100 μF.

The IC we use is the industry workhorse, the LM 741, which is widely available and a really versatile component.

We're going to start off with the simpler circuits and progress up with increasing complexity. There are a lot of components with the later projects, and it is highly desirable to become familiar with these starter circuits. The aim of this book is not just to present a collection of projects but to offer a learning experience as well, to teach you why certain things are the way they are, and explaining how the projects progress from one to another.

Project 1: Inverting Op-amp with Gain ×10

The fundamental building block amplifier, which appears in practically every amplifier design ever published (a slight exaggeration, but not too far off the mark), is the inverting amplifier, so called and recognized by the fact that the input signal is fed to the negative or inverting terminal of the operational amplifier. It is even easier to recognize the inverting amplifier by the presence of a feedback resistor and an input resistor. As the projects progress, you'll see this familiar circuit appearing time and time again.

Circuit Description

Since all the circuits in this book are designed for ac use, they'll have a capacitor feeding the input and another feeding the output. Because this is an inverting op-amp, the input signal will be fed to the negative terminal (that's pin #2). So far we've been able to discern quite a bit about how the circuit should be connected just by looking at the title, "Inverting Amplifier." All inverting op-amps have two resistors for setting the gain—one in the feedback line and the other in the input line. The ratio of the feedback resistor to the input resistor is the gain. For the components values used, this works out to be 100 K/10 K = 10. Thus, if you wanted to increase the gain, the feedback resistor could be increased to, say, 1 Mohm for a gain of 100. With more gain comes more noise and a tendency to overload the following stage. That's why we've stayed with a low but stable value of ×10 here.

If you have access to an oscilloscope you can check that the feedback resistor really provides a ×10 gain. Feed the amplifier with a small amplitude ac signal, from a signal generator if you have one (Project 9 later in this chapter shows you how to build one) to give something like 100 mv. The output will then be 100 mv × 10 = 1 volt.

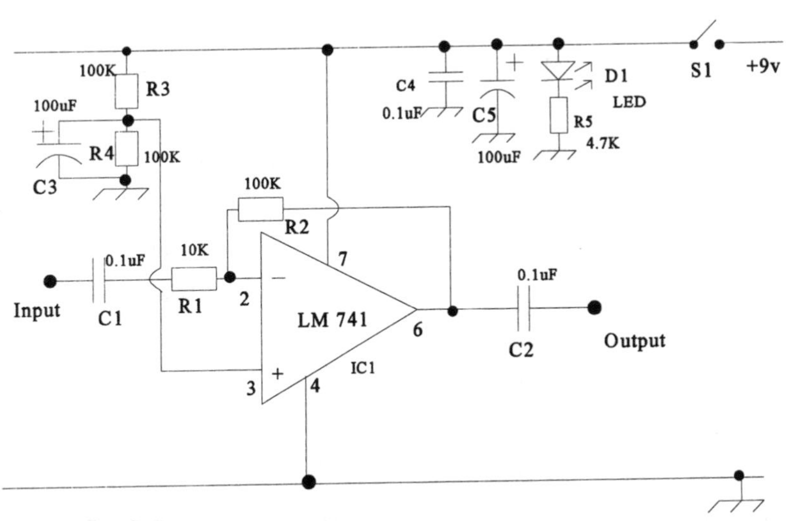

FIGURE 10-1 Inverting amplifier

Back to the circuit, the non-inverting terminal (pin #3) must always be taken to a fixed reference voltage that's half the supply voltage. This is very easily done using two resistors of equal value and a large electrolytic capacitor. Two resistors of equal value placed across the supply will provide a split supply voltage at the junction point. We don't want to draw any current, so the resistors are quite high in value (100 kohms).

A large capacitor stabilizes the split reference voltage but more important it provides a low-impedance ground path for the ac signal. At a frequency of 10 kHz (that's in the audio frequency spectrum), for example, the impedance of a 100 μF capacitor is 0.16 ohm. Therefore, we can call this a short circuit. Two final connections to the IC remain—the very important power (pin #7) and ground (pin #4). That's all there is to it. It's simple once you know what the various components are for. The coupling capacitors are 0.1 μF, and at 10 kHz you get an impedance of 159 ohms, producing a low series impedance.

This circuit is the workhorse of more ac amplifier circuits than you can think of. You can always spot the gain-setting resistors if the input signal is fed to the inverting pin. All the rest of the circuitry does no more than just support the basic amplifying function. Table 10-1 lists some combinations that you can use to vary your gain, based on commonly used resistor values.

TABLE 10-1 Resistor Combinations and Resulting Gain

Gain	*Feedback Resistor*	*Input Resistor*
×1000	1 Mohm	1 kohm
×213	1 Mohm	4.7 kohms
×100	1 Mohm	10 kohms
×21	1 Mohm	47 kohms
×10	1 Mohm	100 kohms
×470	470 kohms	1 kohm
×100	470 kohms	4.7 kohms
×47	470 kohms	10 kohms
×10	470 kohms	47 kohms
×4.7	470 kohms	100 kohms
×100	100 kohms	1 kohm
×21	100 kohms	4.7 kohms
×10	100 kohms	10 kohms

Note: The feedback resistor is usually between 100 kohms and 1 Mohm. The input resistor is usually between 1 kohm and 100 kohms.

Because there's not a lot of drive capability from an LM 741, this type of amplifier is often referred to as a pre-amplifier. For getting an extra boost out of an electric guitar, microphone, or turntable pickup, this is the best circuit to use. A gain of ×10 to ×100 is all that is needed for most applications.

Parts List

Semiconductor

IC1: LM 741

Resistors

All resistors are 5 percent 1/4W
R1: 10 kohms
R2: 100 kohms
R3: 100 kohms
R4: 100 kohms
R5: 4.7 kohms

Capacitors

All non-polarized capacitors disc ceramic
All electrolytic capacitors 25V rating
C1: 0.1 µF
C2: 0.1 µF
C3: 100 µF
C4: 0.1 µF
C5: 100 µF

Additional Components

D1: LED
S1: miniature SPST toggle switch
Power supply: 9-volt battery

Construction Tips

Since this is the first circuit project in the book, take extra care with the connections so that you gain familiarity with the way the components are used. Take note of the polarity requirements on the electrolytic capacitors. Once you've completed the build, you should follow through by going to the next section, "Test Setup," to confirm that your circuit is working.

Test Setup

Apply power to the device and measure the split voltage at the junction of R3 and R4. This should be half the supply voltage. Unfortunately, there's not a lot more you can verify unless you've got a power amp too. If you do have a power amp, you can go ahead with your testing as follows (it's not too critical to test the first two circuits with a power amp because they're given here to illustrate the basic principles of the two fundamental op-amp modes).

First set up a reference condition by connecting a low-level signal source, the simplest would be a microphone hooked into your power amplifier. Keep the volume turned down low so the sound is just barely audible. You need to keep the volume low because you want to verify that the volume gets much louder when the pre-amp is connected, based on what changes you can hear with your ears. Having satisfied yourself that the setup's right, connect up the power amplifier so the pre-amplifier output feeds into the power amplifier's input. This time, the microphone's signal output should definitely be louder. Try experimenting with a radio signal source, too, but remember to keep the signal level low. Provided you've got an increase in signal gain, you can assume that the circuit's working.

Project 2: Non-inverting Op-amp with Gain ×10

This amplifier configuration is much less common than the inverting amplifier, but it does sometimes figure in amplifier circuits you'll run across in magazines and books, so it may help to add this op-amp to your knowledge base. The non-inverting amplifier, as it's name implies, doesn't invert the output signal with respect to the input. But since we cover only ac amplifiers in this book (as opposed to dc amplifiers), that fact makes no difference to us. Because we are making a basic study of audio amplifiers, we're only interested in the non-inverting amplifier's main property of providing amplification or gain, so the treatment here is limited to that property only.

In terms of component count, this op-amp requires an additional capacitor and resistor, over and above the components needed for the inverting amplifier. Purely in terms of gain, going to the non-inverting configuration makes no difference. The distinction between inverting and non-inverting amplifiers only becomes apparent if they were used as dc amplifiers; dc amplifiers, as their name implies, amplifies dc signals. These are not covered in this book. The presence of the coupling capacitors used in ac amplifiers, removes the dc component and allows just the ac signal through. Hence, both types of amplifiers, inverting and non-inverting, perform identically when used as ac amplifiers. As you work through this book and become familiar with the two

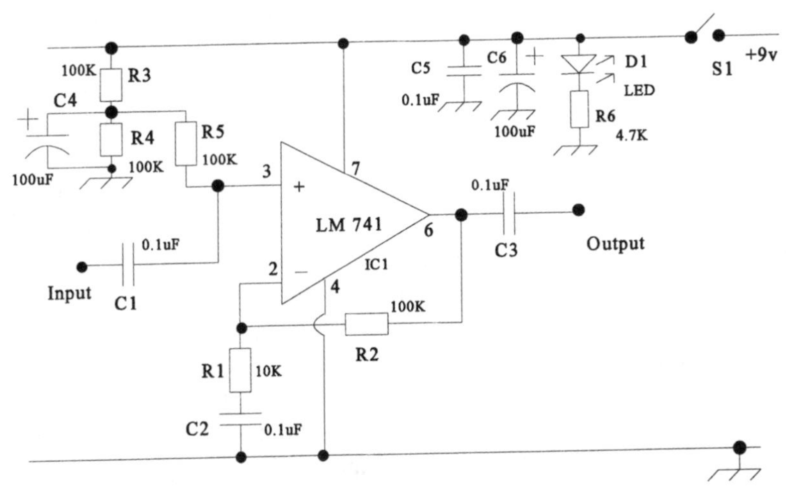

FIGURE 10-2 Non-inverting amplifier

complementary circuits, you'll see instances in which a capacitor is needed to complete the ac path to ground. Here are two distinct examples. The non-inverting amplifier has a capacitor included in the gain-setting network, for completing a low-impedance path to ground; a resistor can't be used for that purpose since a resistor would upset the gain-setting resistors. The other area where a capacitor is needed, in a reverse sense so to speak, is with the split bias supply. In this case the capacitor shunts one of the potential divider resistors to provide a short circuit to ground for the ac signal.

Circuit Description

The non-inverting op-amp can sometimes be found in ac amplifier circuits, but since amplifiers of this sort don't make use of any phase relation requirement between the input and output, there is marginally no difference between the two types—at least from an audio point of view. When we are working with a non-inverting amplifier, we expect the signal to go to the non-inverting positive terminal (to pin #3). If you've already studied or built the inverting amplifier, note the different way in which the representations for the IC symbol are drawn for the non-inverting amplifier. The input signal (a loose convention used here) is always directed toward the upper terminal (verify this for yourself in projects 1 and 2), so the IC input pins are appropriately marked. They are swapped around in the two circuits.

Here's an important point to note: in the non-inverting op-amp the split bias supply always goes to the non-inverting positive terminal. For generating the split supply, we use the same two resistor/one capacitor combination that is used throughout this book, but there's a very important difference here: we must use an additional series resistor so that the ac signal is not shunted to ground. The circuit will not function without this essential component. Very rarely are there any components in ac amplifier circuits that are as critical as this item.

The gain-setting resistor components are, as we might suspect, located between the output and the inverting input (since there's no other input terminal left). The feedback resistor is fine—it's where we expect it to be. But locating the other gain-setting resistor is a little more difficult. There's no signal input here, so where does it go? There's no other place for it to go but ground, and that's where it goes, but first it is connected to a capacitor. Here is another critical point: the ground-return resistor must go through a capacitor when returned to ground. (With all these restrictions to worry about, is it any wonder that the simpler inverting amplifier is more commonly found?) The ground-return capacitor should have a low impedance at the frequency of interest; since we're working in the audio frequency range, our frequency of 10 kHz is going to be our standard "audio frequency." At 10 kHz a 0.1 μF capacitor has an impedance of 159 ohms—practically a short circuit to all intents and purposes. The gain is essentially set by the ratio of the feedback to the return resistor. In the project example, this is 100 kohms/10 kohms or ×10. That's the same value as the coupling capacitor used for bringing the signal into and out of the amplifier. A series impedance of 159 ohms is not going to affect the signal amplitude to any discernible extent. Incidentally, a value of 0.01 μF (1.59-kohm impedance), which is ten times smaller, definitely produces a noticeable attenuation of the signal. So we usually stay at the 0.1 μF value. For a bit more bass (low frequency) boost, an increase to 0.47 μF is fine.

A selection of gains can be produced by selecting the appropriate feedback and ground-return resistor values, as shown in Table 10-2.

TABLE 10-2 Gain and Resistors

Gain	*Feedback Resistor*	*Ground-Return Resistor*
×100	1 Mohm	10 kohms
×10	1 Mohm	100 kohms
×47	470 kohms	10 kohms
×4.7	470 kohms	100 kohms
×10	100 kohms	10 kohms

Parts List

Semiconductor

IC1: LM 741

Resistors

All resistors are 5 percent 1/4W
R1: 10 kohms
R2: 100 kohms
R3: 100 kohms
R4: 100 kohms
R5: 100 kohms
R6: 4.7 kohms

Capacitors

All non-polarized capacitors disc ceramic
All electrolytic capacitors 25V rating
C1: 0.1 μF
C2: 0.1 μF
C3: 0.1 μF
C4: 100 μF
C5: 0.1 μF
C6: 100 μF

Additional Components

D1: LED
S1: miniature SPST toggle switch
Power supply: 9-volt battery

Construction Tips

There are a few more components here as compared with the first inverting op-amp circuit, but work your way through them slowly and you shouldn't go wrong. Take note of the polarity requirements on the electrolytic capacitors, C3 and C5. Once you've completed the build, you need to follow through the next section, "Test Setup," to confirm that your circuit is working.

Test Setup

Apply power to the device and measure the split voltage at the junction of R3 and R4. You should get a reading of half the supply voltage. Unfortunately there's not a lot more you can verify unless you've got a power amp. If you have one at your disposal, you can go ahead as follows (it's not too critical to test using a power amp for the first two circuits since they're given here to illustrate the basic principles of the two fundamental operational amplifier modes). First set up a reference condition by connecting a low-level signal source, the simplest would be a microphone connected into your power amplifier. Keep the volume turned down low so the sound is just barely audible, because you want to verify with your own ears that the volume is much louder when the pre-amp is connected. Having satisfied yourself that the circuit is set up right, connect up the power amplifier so the pre-amplifier output feeds into the power amplifier's input. This time, the microphone's signal output should definitely be louder. Try experimenting with a radio signal source, too, but remember to keep the signal level low. Provided you get an increase in signal gain, you can assume that the circuit's working.

Project 3: Variable Gain Inverting Op-amp

You can transform the basic operational amplifier into an enhanced instrument by adding only one component. The amplifier's gain is controlled in part by the feedback resistor, so vary that and you've got a variation in gain. What could be simpler? Can you spot the extra component? It's the potentiometer, VR1, lurking in the feedback loop, sitting where the normal feedback resistor should be. This is one of the easiest and most useful circuit modifications to make to the basic op-amp configuration. Especially for audio applications, a variable gain option is extremely valuable. You might need the extra flexibility to control gain so it can be maximized without creating distortion. Conventionally, the feedback resistor always has a higher value than the input resistor, so the feedback resistor is the one that we vary in order to provide a variable gain function. Varying the input resistor would also change the input characteristics of the op-amp, but that's something that's not a good design practice. The input resistor can vary over a small range, but generally it stays where it is.

Circuit Description

This circuit offers a variable gain from ×10 to ×100, a range that'll meet the needs of most applications. Because the pre-amplifier is likely to feed into a further power-amplifier stage, we don't want the gain to be too high; otherwise all we'll do is overload the final amplifier and get a distorted output. One of the neat options with this circuit is that it can provide a calibrated label for the 1 Mohm gain control (VR1). Labelling the gain control will add versatility to this project, because then we can gauge the relative sensitivity of various input devices (for example, microphones), by adjusting the gain for a constant output level.

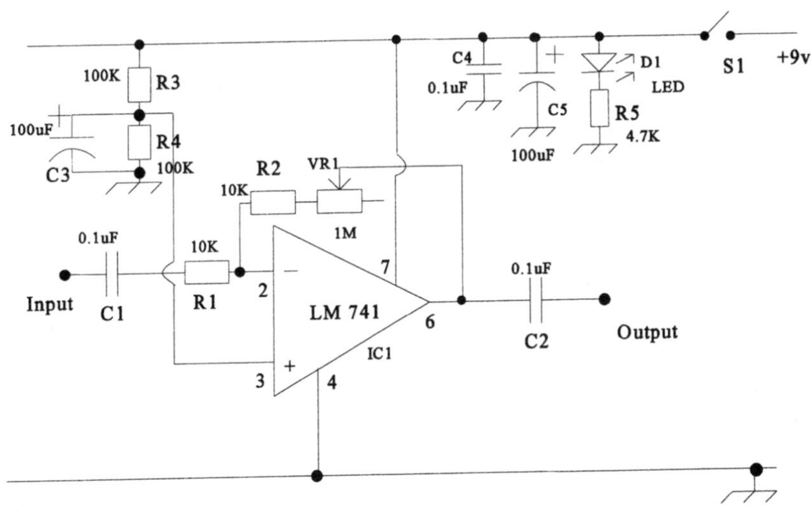

FIGURE 10-3 Variable gain inverting amplifier

There are generally four options for varying gain:

1. The most efficient and best way to vary gain is to vary the feedback resistor, as is done in this circuit.
2. Varying the input resistor has the disadvantage of also varying the input impedance, but this consequence is not desirable from a circuit design point of view, because the input impedance should be a fixed entity.
3. We could maintain the circuit gain high and vary the input signal through a potentiometer. But at a high gain setting the op-amp will contribute more noise, and this noise will be constant regardless of stage gain. This side effect could be undesirable, but, nevertheless, this is a common way of providing volume control at the amplifier stage.
4. Another option for varying gain is to place the volume control at the output end and maintain a constant high gain. Any noise contribution would still be present with this method, so putting the volume control at the output end is not a highly recommended method of controlling stage gain.

Notice in the schematic that there is an extra fixed resistor, R2, connected in series with the variable resistor, VR1 (it's actually just a potentiometer with two terminals used). The purpose of this extra fixed resistor is to prevent the feedback resistance from dropping to zero, as it will do at one end of the potentiometer. The extra series resistor R2 (10 kohms) keeps the minimum feedback resistance to 10 kohms. With the R2 resistor taken together with the component value shown for the input resistor R1, the gain will be 10 kohms/10 kohms = 1. At the other end of VR1's setting, we have a total resistance of about 1 Mohm (ignore the 10-kohm series resistor, since it's small by comparison to 1 Mohm). The gain in this case is 1 Mohm/10 kohms = 100.

Calibrating the Potentiometer

Here's how you want to calibrate the potentiometer (this is done at the final stage of the project, once all the hardware is installed). You need to use a knob with a pointer attached. On the surface of the project case where the potentiometer will be mounted, attach a circular card cutout. Use an ohmmeter to measure the resistance between the free end of R2 and the center wiper terminal of VR1. Mark off the following points on the card to correspond to the following resistance values.

×1	10 kohms
×10	100 kohms
×50	500 kohms
×100	1 Mohm

If there's enough space, fill in the intermediate values to provide greater coverage.

My preference is to do the calibrating once the circuit is known to be working. Switch off the circuit and use a ohmmeter to measure the total resistance across R2 and VR1, and mark off convenient values that are multiples of R1 (10 kohms) on the control knob. For example, R2 + VR1 = 100 kohms, is a gain of 10; R2 + VR1 = 200 kohms is a gain of 20, and so on. Mark in a few of the intermediate values too. The gain is just the ratio of (R2 + VR1) to R1 (which is fixed at 10 kohms). If there's anything connected in parallel with R2 + VR1 that might affect the reading, disconnect that item. To be sure, I generally just de-solder one of the wires to the potentiometer; then I can be absolutely positive that there's nothing connected in parallel.

Parts List

Semiconductor

IC1: LM 741

Resistors

All resistors are 5 percent 1/4W
R1: 10 kohms

R2: 10 kohms
R3: 100 kohms
R4: 100 kohms
R5: 4.7 kohms

Capacitors

All non-polarized capacitors disc ceramic
All electrolytic capacitors 25V rating
C1: 0.1 µF
C2: 0.1 µF
C3: 100 µF
C4: 0.1 µF
C5: 100 µF

Additional Components

VR1: 1 Mohm
D1: LED
S1: miniature SPST toggle switch
Power supply: 9-volt battery

Construction Tips

Whenever a project features mechanical components such as a potentiometer, it's highly recommended that you use a project case, as opposed to laying out the design in a haphazard bird's nest of wires. Mind you, having said that, all the circuits have been built in a deliberately sloppy form, to make sure that no special care is needed. But using a project case helps to prevent wires from accidentally touching and short circuits from occurring, so it's a good practice. It's worth the extra effort. A connection can be made to the potentiometer, VR1, with two flexible wires, that is, the stranded type. Don't use the solid type of hookup wire if you're doing some bench-top work, because a lot of stress will be put on the potentiometer terminals with any sort of mechanical motion, and there is a strong tendency for the solder connections to come adrift. Watch out for the polarity requirements of capacitors C3 and C5.

Test Setup

You'll need some form of power amplifier to drive the output from this project in order to verify that the circuit is operating properly. Any form of signal input can be used. Because of the pre-amp's high-gain capability, a low-level signal source such as a microphone can be used (make sure it's a carbon type that doesn't require a dc bias). With a power amp and speaker connected up, and a signal feed coming in, testing is very simple.

Start with the gain control in the lowest position; the knob should be turned in the full counterclockwise direction. With a signal feeding through, advance the gain control, and the sound output will increase accordingly. The maximum gain setting might cause feedback and howling from the speaker, depending on the gain of the power amp and the proximity of the speaker to the microphone, but we're only interested in making sure the circuit is functioning correctly.

Project 4: Inverting Buffer

There's no other circuit that is likely to garner less interest than the insignificant buffer. With a stifled yawn, I hear you ask, What can it do? It can produce a gain of 1. Isn't that the same as no gain? So why do I need to use it? To really understand the subtleties of the buffer (inverting or otherwise), it's necessary to dig a little deeper into the finer points of circuits; well, more specifically, we need to talk about the interfacing of circuits, or coupling one circuit into another. Simple circuits, as depicted in the earlier sections of this chapter, are invariably stand-alone circuits; that is, there's generally just one IC involved. Yet as we get more ambitious, we find that we can get improved functionality by coupling one circuit into another.

The key word to keep in mind when circuits are coupled together is *impedance,* or, more specifically, *input impedance* and *output impedance*. A typical op-amp circuit, for example, the standard inverting amplifier, has a certain output impedance associated with it. Let's say you have a nice, strong amplified signal at the output. It sounds great, and it has the gain you wanted. The trouble is that the output impedance is fairly high, which in turn means that any low-impedance value load (e.g., a speaker) will start to shunt or reduce the output signal. Thus, the usefulness of an amplifier's output is equated with it's ability to drive a following load. What impedance level would be optimal? If the amplifier's output is low—lower, say, than the load coupled to it—then the load would have a negligible effect on the signal, which is what we want.

The output impedance of the buffer is low. So if the buffer were to feed the low-impedance load, we'd be on the right path. As we've seen earlier, the amplifier's output is affected by low-impedance loads that are coupled to it, so the best type of load (that is, the one that has the least loading effect) would be a high-impedance one. The buffer has a very high input impedance. Suddenly the buffer's ×1 gain seems to be not such a bad idea after all. The buffer is really an impedance convertor: it converts a high-impedance source into a low-impedance source so that you can feed the buffered signal into a low-impedance load. The buffer by itself doesn't have the ability to drive a lot of current, which means that it cannot sink huge amounts of currents, but for driving meters it's more than adequate.

It is important to note that if you just wanted a ×1 gain, then the gain-setting resistors in a regular non-inverting op-amp could be set to the same value, but in that case you would not get the high input impedance and low output impedance that you get with a buffer.

Circuit Description

What's a buffer? It is a unity gain amplifier; that is, it produces a gain of ×1. Where the source already has a high enough signal level, you don't want to add any more gain, hence the unity gain setup.

In the buffer circuit there's even an attenuator at the input to reduce high-level inputs, for example, that coming from the earphone output of a radio. The input is fed through a simple two-resistor potential divider. The degree of attenuation is determined by the ratio R2/(R1 + R2), which in this case is

10 kohms/(27 kohms + 10 kohms) = 10/37 = 0.27 kohms

or about a reduction of a third. This is appropriate when you are experimenting with a radio signal output to prevent overloading to an external power amplifier. Don't forget the radio signal coming out of the earphone socket is already at a high-voltage level, hence the need for the attenuation. Make sure you place the attenuator circuit before the input capacitor (C1), as shown in Figure 10-4. It's quite easy to make a wiring error on the circuit board and place the attenuator circuit after C1, so be careful here.

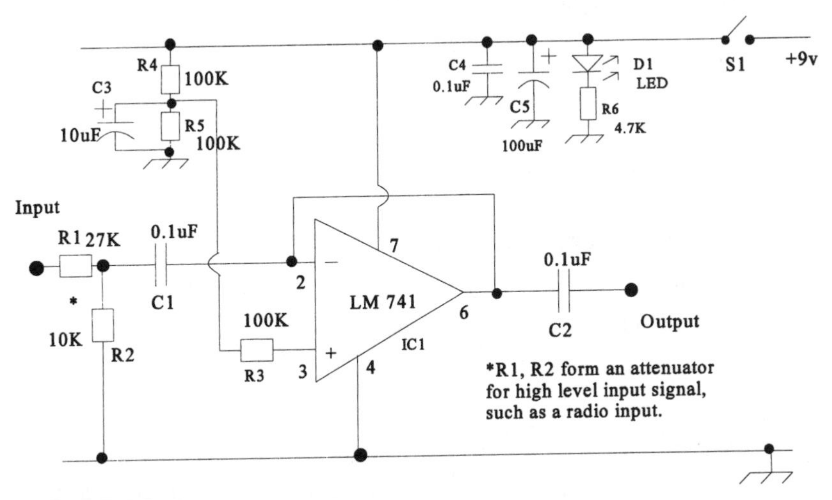

FIGURE 10-4 Inverting buffer

As the name implies, we are building an inverting buffer, so the input signal goes to the negative terminal (pin #2) via the customary input capacitor (C1). But if you're expecting to see the usual feedback resistor/input resistor combination, it's not here. Granted, the gain could be made to unity if we had the two resistors and these were made equal, but the way we connect things here is even simpler. The output signal from pin #6 is connected directly to pin #2 (the negative input). This circuit configuration has a high impedance, so it doesn't load the signal feeding into it. It has a low impedance, which facilitates driving the output into a following stage.

Parts List

Semiconductor

IC1: LM 741

Resistors

All resistors are 5 percent 1/4W
R1: 27 kohms
R2: 10 kohms
R3: 100 kohms
R4: 100 kohms
R5: 100 kohms
R6: 4.7 kohms

Capacitors

All non-polarized capacitors disc ceramic
All electrolytic capacitors 25V rating
C1: 0.1 μF
C2: 0.1 μF
C3: 10 μF
C4: 0.1 μF
C5: 100 μF

Additional Components

D1: LED
S1: miniature SPST toggle switch
Power supply: 9-volt battery

Construction Tips

There aren't too many components for this circuit, so you should be able to build it in one go. It's a good idea to use IC sockets because they make IC replacement so much easier. Watch out for the polarity requirements of the electrolytic capacitors, C3 and C5. Don't forget resistor R3; it has to be there for the circuit to function properly. There's a little extra attenuator circuit at the front end. I've added it because I often use a radio signal (taken from the earpiece socket) as a signal source, and these two components (R1, R2), cut down the high-level signal of the radio by just the right amount.

Test Setup

The test setup for the buffer circuit is somewhat different from the tests for most of the other circuit projects whose functions are fairly obvious. It takes a little more cunning to evaluate the buffer, mainly because of its very passive or inert nature. You'll need to feed in a signal to start with. With the attenuator at the input, a simple radio signal source can be used. The buffer's output needs to be monitored; it could be a regular power amplifier feeding a speaker.

Once the setup is running, you can place a variable resistor between the buffer output and ground. As the shunt resistance is reduced (use a starting value of 100 kohms), you can detect the point at which the signal becomes noticeably reduced. The shunt resistance can then be measured to give you a reference point for the buffer. Later on if you come to build the regular non-inverting amplifier, repeat the experiment again and you'll find that the shunt resistance will be very much higher, showing that the buffer allows a lower impedance load to be driven.

Project 5: Non-inverting Buffer

Under ac operating conditions there is no distinction between op-amp-based circuits configured in the inverting (more common) or non-inverting mode. But to be comprehensive, I present the non-inverting buffer as well.

Buffers show off their finer points when they are used as an interface between two circuits; on their own they really don't generate a lot of excitement. The buffer's gain is very mundane: what could be more uneventful than a circuit with a gain of ×1. That means the signal that comes out is exactly the same as the signal that went in. Regular amplifiers are good because they boost the incoming signal, but what's good about a ×1 amplifier? The key to understanding the benefits of the buffer can be summed up in the word *matching*.

Circuits in general and amplifiers in particular are rarely used in a stand-alone situation; there is inevitably a string of following circuit elements. To match a pre-amplifier to a following stage, let's say a low-impedance load such as a speaker, it is critical that the load does not shunt the pre-amplifier's signal. This possible loading effect is controlled by the pre-amplifier's output impedance. As a general rule of thumb, if the output impedance of a circuit is feeding into a following stage with a lower impedance, the signal is liable to be reduced as a result of the loading effect. The severity of the problem is a factor of the relative values of the two impedances. To nullify or negate this undesirable effect, we need something that will act as a buffer; we need something with a high input impedance so that the incoming signal won't be loaded, and a low output impedance that will allow a feed into a low-impedance load. The ×1 non-inverting buffer amplifier has exactly those properties and solves the problem. Where the final load impedance is not too low, say around a kohm or so, the simple op-amp buffer shown here is ideal.

It is important to note that if you just wanted a ×1 gain, then the gain-setting resistors in a regular non-inverting operational amplifier could be set to the same value, but you would not get the high input impedance and low output impedance associated with using a buffer.

Circuit Description

The non-inverting buffer provides no phase inversion between the output and input signal, although for ac audio applications this factor has no special relevance. A buffer provides very good isolation between two signal stages, and, in keeping with its isolating properties, it has a unity gain that provides further neutrality. The high input impedance that is characteristic of buffers inhibits any loading effect on the input signal. The buffer's low output impedance facilitates the connection of a following stage, and we don't have to worry about the latter stage shunting the buffer's output signal.

The signal input goes to the positive terminal (pin #3), and in keeping with the requirements for a single-supply design, the positive terminal requires a split supply bias for correct operation. An isolation resistor (R5) provides the coupling to the correct bias voltage. Since the application is designed for coupling a radio receiver output into an external power amplifier, there is a very necessary passive attenuator network at the input. This is just a simple but efficient two-resistor (R1 and R2) potential divider, producing an attenuation of about 30 percent with the component values shown. Incidentally, inexpensive piano-key-type cassette recorders (mono versions) work well with this buffer acting as an interface between the cassette recorder and a higher performance amplifier.

All buffer circuits produce unity gain (×1), but that's not a problem, because they are usually used as interface devices, and the necessary amplification is produced by the source. Make sure you place the attenuator circuit before the input capacitor (C1), as shown in Figure 10-5. It's quite easy to make a wiring error on the circuit board and place the attenuator circuit after C1, so be careful here.

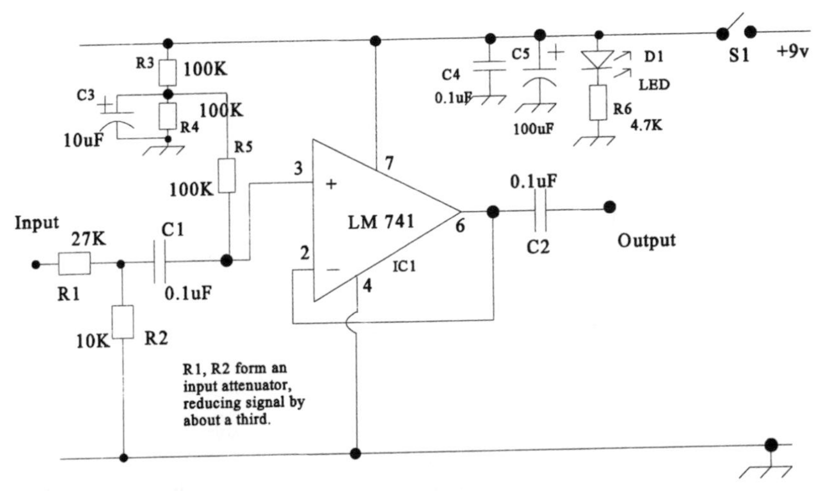

FIGURE 10-5 Non-inverting buffer

Parts List

Semiconductor

IC1: LM 741

Resistors

All resistors are 5 percent 1/4W
R1: 27 kohms
R2: 10 kohms
R3: 100 kohms
R4: 100 kohms
R5: 100 kohms
R6: 4.7 kohms

Capacitors

All non-polarized capacitors disc ceramic
All electrolytic capacitors 25V rating
C1: 0.1 μF
C2: 0.1 μF
C3: 10 μF
C4: 0.1 μF
C5: 100 μF

Additional Components

D1: LED
Power supply: 9-volt battery

Construction Tips

There aren't too many components for this circuit, so you should be able to build it in one go. It is a good idea to use IC sockets because they make IC replacement so much easier. Watch out for the polarity requirements of the electrolytic capacitors, C3 and C5. Don't forget resistor R5, and make sure it's coupled to the correct side of capacitor C1. The front-end attenuator is optimized for feeding a radio signal taken from the earpiece socket.

Test Setup

The test setup for the buffer circuit is somewhat different from the tests for most of the other circuit projects whose functions are fairly obvious. It takes a little more cunning to evaluate the buffer circuit, mainly because of its very passive or inert nature. You'll need to feed in a signal to start with. With the attenuator at the input, a simple radio signal source can be used. The buffer's output needs to be monitored; it could be a regular power amplifier feeding a speaker.

Once the setup is running, you can place a variable resistor between the buffer output and ground. As the shunt resistance is reduced (use a starting value of 100 kohms), you can detect the point at which the signal becomes noticeably reduced. You can measure the shunt resistance to get a reference point for the buffer. Later on if you come to build the regular non-inverting amplifier, repeat the experiment again and you'll find that the shunt resistance will be very much higher, showing that the buffer allows a lower impedance load to be driven.

Project 6: High Gain Inverting Amplifier with High-Frequency Cut Filter

Op-amp gains can be easily increased to startlingly high values by merely raising the feedback resistor and lowering the input resistor. At some point, however, the circuit could become unstable and oscillate because of the high gain, or too much noise may be generated. Adding a low-value shunt capacitor across the feedback resistor is a simple and effective technique for controlling the amplifier's stability, and it also cuts back on high-frequency (hf) hiss. The circuit shown here effectively demonstrates this common technique, which is employed in many circuits you'll come across in other books and magazines. Later on you'll find that this capacitor is a very simple yet effective means to tailor high-frequency response, but for now we'll start with its use as a basic hf cut device. This circuit also makes volume control possible, to add a little variety to the standard setup. You can use a variety of both low-level and high-level signal sources. These components apart, it's a fairly standard circuit. The stage gain is exceedingly high

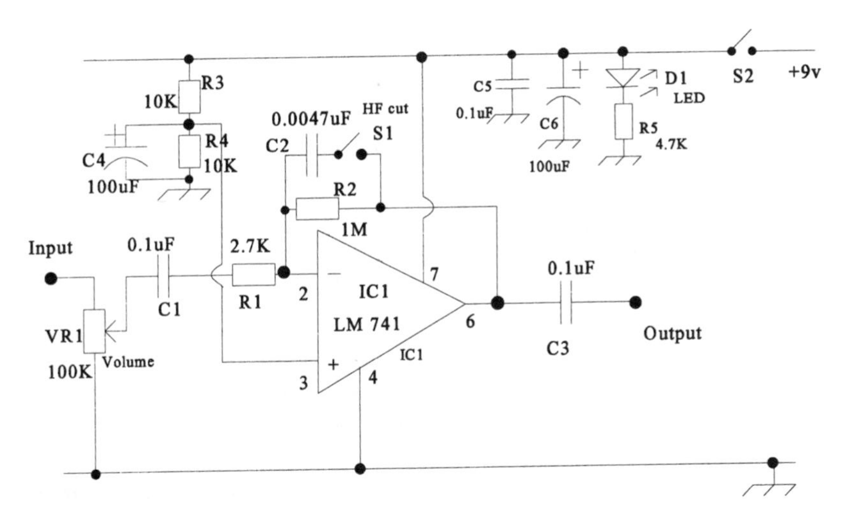

FIGURE 10-6 High gain inverting amplifier with high-frequency cut

because of the very high value of the feedback resistor used (1 Mohm). That particular resistor (R2) provides a little spice to the different circuits and demonstrates the range of component values that can be used.

Circuit Description

The inverting op-amp configuration presented here is more complex than the one we've seen before. It has a very high gain of ×370, determined by the ratio of R2 to R1 (1 Mohm/2.7 kohms). At such high gains the amplifier is likely to overload and go into distortion, so an input potentiometer (VR1) is also included in the circuit to limit the incoming signal. There is no limit to the high-frequency response using just R2 and R1, and therefore a shunt capacitor (C2) is included in a switchable format using switch S1. Capacitor C1 (0.0047 µF) causes the amplifier gain to fall off (reduce) dramatically. At 10 kHz the capacitive impedance is 3.39 kohms. In parallel with the 1 Mohm resistance of R2, the resulting resistance is 3.38 kohms—practically the same value as the capacitive impedance. When two impedances are connected in parallel (which is what is done here), the effective resistance is governed by the smaller value, if there is a large difference between the two values. At the lower frequency of 1 kHz, C1's impedance increases to 33.9 kohms, and at 100 Hz the impedance climbs to 339 kohms. Therefore, we can see that the gain (with C1 switched in) falls as the frequency is increased, thus producing a more mellow high-frequency cut. Of course, the maximum gain has fallen drastically from the dc value of ×370, but in reality it is unlikely that the full value will be used, so the absolute dc value makes little difference.

The input potentiometer (volume control) allows the circuit to accommodate a wide range of signal levels. Notice that the potentiometer comes before the input capacitor C1. The circuit will not function if you should inadvertently swap the positions of VR1 and C1. This is quite easily done at the assembly stage. Table 10-3 shows the effect of the shunt capacitor on the calculated gain.

If you want even more high-frequency cut, try a value two times larger for C1, for example, 0.01 µF. In Table 10-4 you can see the effects: the gain is correspondingly reduced.

What's interesting is to see how the gain falls off with frequency with C1 = 0.0047 µF and C1 = 0.01 µF. To do that we first normalize the gain to the highest value; in this case we normalize the gain at 100 Hz. This means we divide this number into all the other gain numbers. This calculation clearly brings out the gain-frequency dependence. The ratio is multiplied by 100 to convert it to a percentage.

TABLE 10-3 0.0047 µF Shunt Capacitor and Calculated Gain

C1 = 0.0047 µF	*f = 100 Hz*	*f = 1 kHz*	*f = 10 kHz*
Impedance of C1	339 kohms	33.9 kohms	3.39 kohms
Impedance of C1 in parallel with 1 Mohm	253 kohms	32.79 kohms	3.38 kohms
Overall gain	253/2.7 = 93.7	32.79/2.7 = 12.14	3.38/2.7 = 1.49

TABLE 10-4 0.01 µF Shunt Capacitor and Calculated Gain

C1 = 0.01 µF	*f = 100 Hz*	*f = 1 kHz*	*f = 10 kHz*
Impedance of C1	159.13 kohms	15.9 kohms	1.59 kohms
Impedance of C1 in parallel with 1 Mohm	137 kohms	15.65 kohms	1.16 kohms
Overall gain	137/2.7 = 50.7	15.65/2.7 = 5.8	1.16/2.7 = 0.43

TABLE 10-5 Normalized Gain Values

	f = 100 Hz	*f = 1 kHz*	*f = 10 kHz*
C1 = 0.0047 µF: Normalized gain	93.7/93.7 = 100%	12.14/93.7 = 13%	1.49/93.7 = 1.6%
C1 = 0.01µF: Normalized gain	50.7/50.7 = 100%	5.8/50.7 = 11.4%	0.43/50.7 = 0.85%

Parts List

Semiconductor

IC1: LM 741

Resistors

All resistors are 5 percent 1/4W
R1: 2.7 kohms
R2: 1 Mohm
R3: 10 kohms
R4: 10 kohms
R5: 4.7 kohms

Capacitors

All non-polarized capacitors disc ceramic
All electrolytic capacitors 25V rating
C1: 0.1 µF
C2: 0.0047 µF
C3: 0.1 µF
C4: 100 µF
C5: 0.1 µF
C6: 100 µF

Additional Components

VR1: 100 kohm potentiometer
D1: LED
S1: miniature SPST toggle switch
S2: miniature SPST toggle switch
Power supply: 9-volt battery

Construction Tips

The high-gain inverting amplifier circuit is relatively simple to construct and is an obvious extension of the basic inverting amplifier. Because of the presence of the input volume control (VR1) and the filter switch (S1) it is a good idea to use a project case for these mechanical components. I've always found when experimenting with prototype layouts that mechanical components are almost impossible to manipulate with one hand, and there's also a risk of wires moving to places they shouldn't be and creating shorts.

Test Setup

To hear the impact of the high-frequency filter, it is essential to feed this input into a power amplifier. The signal should be a low-level source because the stage gain will be set to a high maximum value. Use the input volume control (VR1) to cut back on any signal clipping or induced distortion. You should be able to discern a very noticeable cutoff of the higher frequencies as C2 is switched in, and the signal should become quieter. The smaller the value for C2, the less the high-frequency cutoff effect. To reduce the high-frequency signal even more, C2 is increased. A typical working range for C2 is anywhere between a few 100 pF and 0.01 µF. The actual value will depend on the value of the feedback resistor used, because it is the parallel impedance combination of C2 and R2 that determines the high-frequency cut-off point.

Project 7: Pre-amp with Bass-Treble Control

Most of the basic amplifier configurations using op-amps have a fairly limited frequency response. Generally this is not much of a concern since the LM 741 is not really an audio quality IC. But by beefing up the input circuitry you can improve the response dramatically. Your sound quality will be distinctly superior to that of a "regular" amplifier with this active tone control circuit. If you use a good quality, large-diameter (8-inch) speaker, you can improve the sound quality of "tinny" sound sources significantly by adding an improved bass bite and dynamic sharp treble bites. The smarts are taken care of by separate treble and bass frequency processing networks. The signal is attenuated, in the process of being enhanced, but the attenuation is easily taken care of by adding some post-processing amplification. You can emphasize high frequency (treble) by accentuating the high frequencies at the expense of the middle and bass frequencies. Similarly the bass frequency is emphasized by accentuating the low frequencies at the expense of the middle and high frequencies. This distinction is important because the simple tone control often found on radio sets merely cuts down on the high frequencies to simulate a basslike effect. The results are not the same as those to be had with a proper bass/treble circuit.

The component count in this design is high, and the circuit looks a little component heavy at first glance, but I'll take you through it slowly in the "Circuit Description" section.

Circuit Description

At first glance this circuit looks a bit daunting. But if you've followed the earlier projects, it's not that complex. First of all, let's get rid of

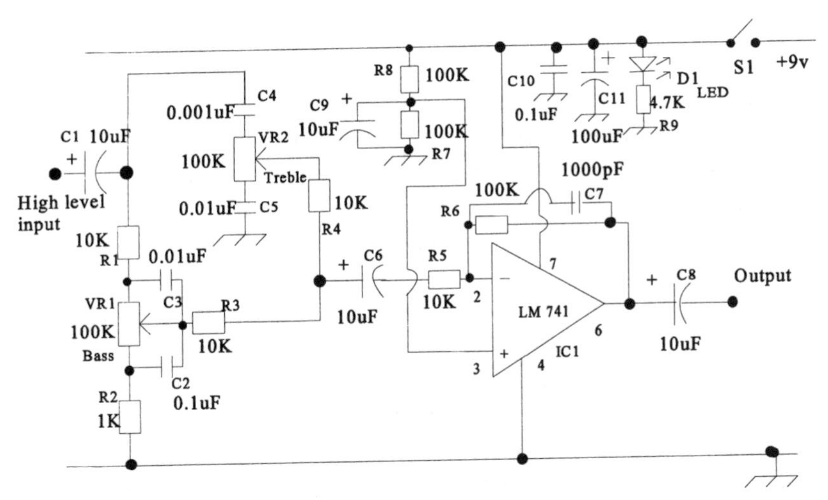

FIGURE 10-7 Pre-amplifier with bass-treble control

the usual supporting components, such as the switch, the LED, the decoupling capacitors, and the split bias supply network. Now what's left? The op-amp is an inverting amplifier with a gain ×10 (100 kohms/10 kohms). See how easy it is! There's a small capacitor (100 pF) across the feedback resistor, so there's a little high-frequency cut applied here—not a lot, given that C7 is relatively small. There are the usual input and output coupling capacitors, but because we will be wanting to emphasize the lower frequencies, the coupling capacitors are increased in value to 10 µF (from the customary 0.1 µF). At the low frequency of 100 Hz, that means a change of impedance from 15.9 kohms to 159 ohms, when we go up in capacitance from 0.1 µF to 10 µF. That's a significant change, and, more important, it means an increase in the bass frequencies. Note the polarity requirements for the two capacitors, C6 and C8.

The bass-treble networks are shown to the left of the IC. The bass-treble network is drawn conveniently separated out for easy identification. Many other publications often show the exact same circuit drawn with the components so interleaved it's impossible to separate out the bass and treble halves. Most likely these circuits are drawn by a draftsperson rather than an electrical engineer, and with repeated copying from book to book, the confusion remains. (Incidentally, all the circuits here are originally-drawn by the author and are not to be found in any other book. My publisher transcribes the layout exactly as I send it to them [thanks Pam].)

This particular circuit has the added benefit of being totally independent of the op-amp circuit. Each section (the treble or bass) can be independently extracted, and the circuit will still function as a regular op-amp.

The bass network is made up of the components R1, VR1, R2, C3, C2, and R3. For the time being ignore the capacitors and resistor R3, which merely acts as an isolating resistor. The remaining components—R1, VR1, and R2—look just like a potentiometer. Remove the fixed resistors (R1 and R2) and you're left with just a potentiometer. Resistors R2 and R1 have been determined through previous experimenters to be the values shown (see Figure 10-7), along with the value for VR1. So the whole package should stay together—you shouldn't vary any of the components.

Having seen that we have a very basic potential divider circuit, now let's add back capacitors C2 and C3. These serve to tailor the response such that the bass frequencies are boosted when VR1 is shifted to one side of the center position, and the opposite happens when the wiper goes the other way: bass frequencies are cut. This circuit is a nice, simple, elegant circuit. The signal is attenuated as a result of the introduction of this passive network; that is, the value of the signal when it exits (albeit with bass boosted/cut) is less than its starting value. To compensate, there's also a ×10 gain pre-amplifier.

Up at the top end of the circuit, there's the equivalent treble control. The components are C4, VR2, C5, as well as the same isolating resistor, R4. Once again, stripped down to just the bare components, we see the signal feeding a potentiometer (VR2) and the output taken (as we would expect) from the center terminal. When we add capacitors C4 and C5, the response changes dramatically. To one side of the center position the treble is enhanced, and moving in the opposite direction the treble is cut. While a treble cut masks the high frequencies, it is not the same as adding bass boost. You really need the bass network for that.

To make full use of this active tone control network, it must be fed into an appropriate power amplifier and speaker combination that is able to do justice to the boosted frequencies. It's no use using a tiny transistor radio replacement speaker and hoping to hear bass. Use a good-size speaker, at least 8 inches in diameter, with an extended response. Similarly, a tweeter would help open up the treble notes. The input signal is commonly coupled to both bass and treble networks through C1 and resistors R3; R4 prevents unwanted interaction between the two circuits.

The combined bass-altered and treble-altered signal is fed via capacitor C6 to our usual inverting op-amp. The small-value capacitor, C7, connected across the feedback resistor R6 cuts down some of the higher-frequency components.

Parts List

Semiconductor

IC1: LM 741 op-amp

Resistors

All resistors are 5 percent 1/4W
R1: 10 kohms
R2: 1 kohm
R3: 10 kohms
R4: 10 kohms
R5: 10 kohms
R6: 100 kohms
R7: 100 kohms
R8: 100 kohms
R9: 4.7 kohms

Capacitors

All non-polarized capacitors disc ceramic
All electrolytic capacitors 25V rating
C1: 10 µF
C2: 0.1 µF
C3: 0.01 µF
C4: 0.001 µF
C5: 0.01 µF
C6: 10 µF
C7: 1000 pF
C8: 10 µF
C9: 10 µF
C10: 0.1 µF
C11: 100 µF

Additional Materials

VR1: 100-kohm potentiometer
VR2: 100-kohm potentiometer
D1: LED
Power supply: 9-volt battery

Construction Tips

It's best to build this project in stages, verifying each as you go along. Verify the basic op-amp stage first using R5, R6, C6, C8, and the split bias components R7, R8, and C10. The gain is ×10 (100 kohms/10 kohms), and the output can be checked using any method. If you're new to this, the output needs to be fed to a following power amplifier and speaker combination because IC1's output by itself cannot drive a low-impedance load (such as a speaker). A signal source such as a cassette or a radio input can be used. Once you're satisfied all is well, switch off the power, remove the temporary connection, and continue with the build.

Check out the treble section next. You should be getting a treble boost when VR1 is in the fully clockwise position, and a treble cut when VR1 is in the fully counterclockwise direction. If your results are the opposite, just reverse the connections to the two outer terminals of VR1. Treble cut might sound like bass boost, but this is actually not what is happening. When treble is cut, the high frequencies are cut, giving you a smoother sound. The cymbals on a drummer's kit will be masked by a treble cut setting, so for that you want the treble boost position.

Once the treble is fine, go on to check out the bass section. Similarly, there should be bass boost when VR2 is in the fully clockwise direction, and bass cut when VR2 is in the fully counterclockwise direction. Bass cut will give you a flat sound, but the more popular bass boost position adds a lot more life to blues tracks. The signal output from a mono cassette recorder (piano-key style) is essentially devoid of any bass response. Feeding such a signal into this circuit will provide a noticeable improvement in sound quality, provided that you have a decent-size, properly mounted speaker.

Test Setup

To properly appreciate the capabilities of this impressive circuit, use a power amplifier for the final circuit, together with a good quality speaker (properly mounted in a speaker cabinet) that is capable of reproducing bass and treble frequencies. Depending on the signal input used, that is, whether that input inherently has a wide-frequency response, you'll find that with the tone controls set to the full bass and full treble positions, the sound you get will definitely have an edge over the plain untreated sound. I've used this circuit to add considerable bass/treble enhancements to a very plain mono signal from an inexpensive piano-key-type cassette recorder. This type of cassette recorder produces a much plainer signal (it's practically devoid of any bass enhancements, since its primary purpose is for recording voice) than that of the high-fidelity Walkman type of cassette recorder (which is specifically designed for high-quality sound outputs).

The components around the bass and treble networks go together as a set and should not be varied one at a time, unlike the simpler components around the basic op-amp (which can be varied one at a time). This bass/treble network is of a particularly nice design (there are other variations on the same theme), because it is independent from the active op-amplifier

Project 8: Power Amp with Gain and Bass Boost

The deceptively tiny 8-pin LM 386 IC can be turned into a very passable quality power amplifier that is capable of providing more than adequate amounts of listening power. By making use of the device's internal capability to add extra gain and boost the bass frequencies you can get a very good quality sound. Using a good quality speaker around 8 inches in diameter, with a good frequency response and mounted in a proper enclosure, you'll be able to make good use of the bass boost feature with this device. With this circuit there is more than enough power to suit most listening requirements. With a sufficiently large signal fed into the input it is unlikely that the amplifier will be operated at full volume. A passive tone control adds a little flexibility to your ability to cut down the high frequencies, to provide a less harsh sound. Switching options for the added features (high gain and bass boost) allow you even more flexibility to tailor the sound to suit your own tastes.

Circuit Description

The principal distinction separating a power amplifier from a pre-amplifier is the power amplifier's ability to drive a low-impedance load or more commonly a loudspeaker or headphones. A loudspeaker has a low impedance, typically in the 8-ohm region, so a power amplifier is needed to provide the drive capability. Regardless of whether the output power is minuscule (less than 1 watt for small transistor radios), current still needs to

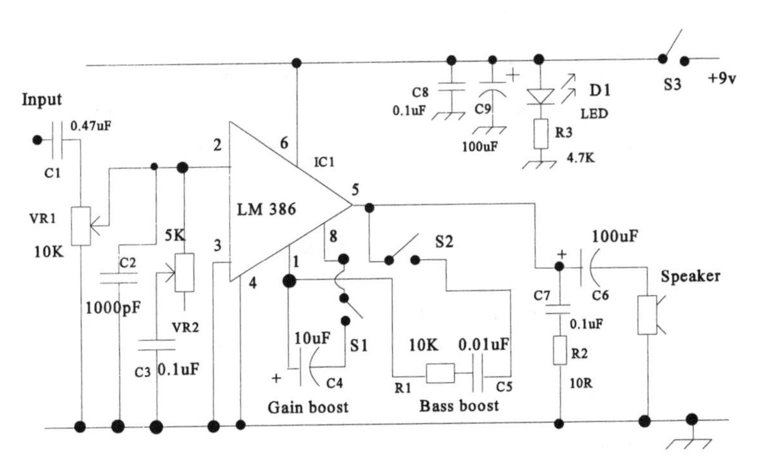

FIGURE 10-8 Power amplifier with gain boost, bass boost

be fed through a low-impedance load. Thanks to a very versatile semiconductor, the LM 386, the tedious task of putting together a power amplifier is made much more pleasant.

The circuit described here is very comprehensive and offers an input volume control to cater to various input signal levels, and a simple but effective tone control to effectively shunt the higher frequencies. In its minimum gain mode, the LM 386 has a ×20 gain, which is more than adequate for high-level signals. For weaker signals, the gain can be boosted significantly to ×180 by simply connecting a capacitor, C4, between two terminal pins (#1 and #8). There is the expected increase in noise at the higher gain levels. With this versatile IC you also have the option to extend the bass response by connecting a series resistor-capacitor combination, R1, C5, between pins #1 and #5. Control of the input signal is through the potentiometer, VR1. The input signal is coupled through capacitor C1, and the wiper of VR1 goes directly to pin #2. The wiper feed can be coupled with either dc or ac. From pin #2 there is a small capacitor, C2, to ground. This capacitor serves to decouple any radio interference pickup especially at the high gain setting. Also connected at pin #2 is the VR2, C3 combination to ground. As VR2's resistance is decreased, more of the high frequency is shunted to ground via C3. At 10 kHz, C3 has an impedance of only 159 ohms.

As mentioned earlier, this circuit has a gain boost and bass boost capability added in. Two switches, S1 and S2, allow these functions to be switched at will to suit a variety of individual requirements. Note the polarity requirement for C4. The audio power output is taken from pin #5 through a large electrolytic capacitor, C6 (100 μF), to make sure the bass-boosted frequencies are not attenuated. A resistor-capacitor combination, C7 and R2 (called a Zobel network), shunts the output signal before C6 to ground. This helps to ensure a cleaner output at higher volume levels. To make use of these enhanced frequencies, the speaker should be a high-quality, wide-frequency response type, preferably at least 8 inches in diameter and housed in a proper enclosure. The power output level is just under 1 watt and is more than adequate for most experimental applications. Current consumption would depend on the volume level. Generally, the current usage goes up as the power level is increased. Under quiescent conditions, however, when there is no input signal, only a few milliamps of current is drawn. Any of the pre-amplifier circuits can be coupled into this versatile power amplifier.

Parts List

Semiconductor

IC1: LM 386 audio power amplifier

Resistors

All resistors are 5 percent 1/4W
R1: 10 kohms
R2: 10 ohms
R3: 4.7 kohms

Capacitors

All non-polarized capacitors disc ceramic
All electrolytic capacitors 25V rating
C1: 0.47 µF
C2: 1000 pF
C3: 0.1 µF
C4: 10 µF
C5: 0.01 µF
C6: 100 µF
C7: 0.1 µF
C8: 0.1 µF
C9: 100 µF

Additional Materials

VR1: 10-kohm potentiometer
VR2: 5-kohm potentiometer
D1: LED
S1: SPST miniature switch
S2: SPST miniature switch
S3: SPST miniature switch
Power supply: 9-volt battery

Construction Tips

Begin with a basic version of the power amplifier and use an IC socket for whatever assembly board you're using. As a temporary measure leave out all of the components (VR1, C1, C2, VR2, C3) around input pin #2 and just feed in a signal via a capacitor with a value around 0.1 µF. For now, also leave out the gain boost (C4), bass boost (R1, C5), and Zobel network (C7, R2). As you can see if you've worked with the LM 386 before, the circuit is reduced to just the basic ×20 gain amplifier. Hook up a speaker or headphones to the output (don't leave out C6) and feed in a fairly high level signal (for example, a radio signal from the earphone socket). All being well, you should get a strong, healthy signal. On that basis, add back the rest of the components and build the circuit. Take care with reading the resistor color codes. The resistor R2 is a particularly low value, and if it is mistakenly swapped with something else, it's very likely the circuit won't function. I always check all resistors with an ohmmeter just to make sure I've read the color code correctly. Sometimes distinguishing the correct colors becomes a little tricky, but the ohmmeter will sort out any doubts.

Test Setup

Hook up a good quality, decent-size speaker (housed in a proper enclosure) to the output and set up a signal source, perhaps a cassette recorder input. Turn on the power to the amplifier first before applying a signal input so as not to damage the circuit. Leave the gain switch (S1) in the open position, and do the same with the bass boost switch (S2). Set tone control VR2 to the midpoint. Set VR1 also to the midpoint. When you apply a signal you should get a good signal from the speaker. Adjust your signal input to a low to medium level to begin with. With all being well, check that the gain in increased as VR1 is rotated clockwise. Next check the tone control, VR2. As VR2 is rotated clockwise, the higher frequencies should be increasingly attenuated, until when the dial is at the full clockwise position, the signal should definitely be more mellow sounding. With the volume kept fairly low, switch in the gain boost (S1). Since the gain will have increased from ×20 to ×180, the signal level will jump significantly and possibly even distort. Reduce VR1 to get a comfortable level. Now check out the bass boost circuit. When S2 is switched in, the bass frequencies will be very strong. Your speaker must have the frequency range to optimize this low-frequency enhancement. The speaker should be a high quality, high fidelity speaker at least 8 inches in diameter with a listed frequency response extending to somewhere below a 100 Hz; the lower, the better.

Project 9: Audio Signal Generator

Sooner or later (probably sooner) you're going to need an audio signal source to verify that pre-amplifiers, power amplifiers, speakers, and headphones are working properly. A true sine-wave generator presents too many construction problems, and in any case it is not necessary for what we have in mind. For the record, is there an equally simple, cheap, readily available method for generating a sine wave with adequate drive capability to match the simple circuit shown here? I think not. There are specialist ICs that produce a clean sine wave, but these are unlikely to be found at your local components store. Remember, one of the philosophies of this book is to use only readily available components that anyone anywhere in the United States has

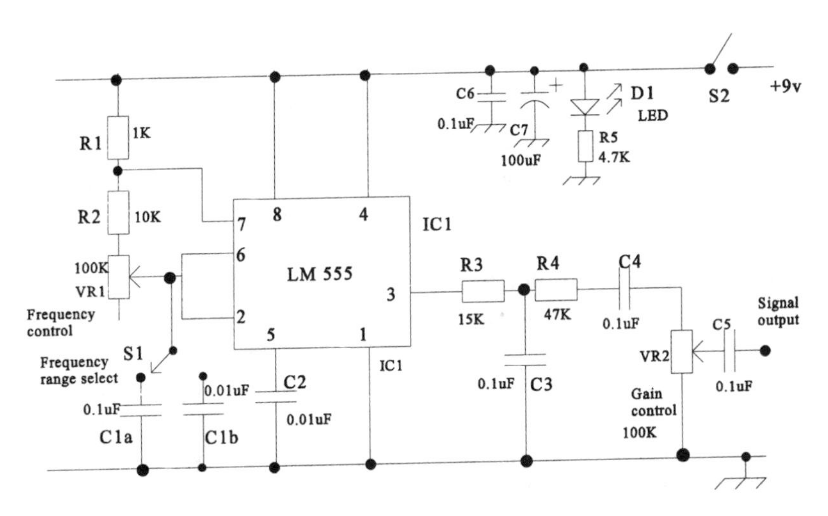

FIGURE 10-9 Audio signal generator

access to. And there are really only four ICs that fall into that category, in my opinion: the LM 741 (single IC op-amp), the LM 555 (timer IC), the LM 386 (audio power amplifier IC), and the LM 324 (quad op-amp IC).

This audio signal generator project is so versatile I'm betting you'll put it into service immediately as soon as the switch is thrown. For added versatility, rather than generating only a straight signal source, there are options for making frequency adjustments and for using a frequency range selector and output amplitude control. When this audio signal source is mounted inside a nice project case, you'll have one of the most widely used pieces of test equipment on your bench. Even if you have a commercial signal generator, the neat little unit you'll build here will be much easier to use.

Circuit Description

Soon after you've built the audio amplifiers, you'll be requiring the use of a test signal source. Using the versatile LM 555 timer IC, you can build a really useful audio signal generator that will prove invaluable for checking out all sorts of audio devices. The LM 555 is an 8-pin IC in a dual-in-line (DIL) package.

In the free-running or astable mode, a square wave generator can be constructed from just a handful of components. Although the duty cycle (the ratio of the on period to the total cycle period) varies when the frequency is varied, this is not a serious issue. A variable resistor, VR1, provides the ability to continuously adjust frequency over a specific bandwidth that is determined by the selection of either timing capacitor C1a or C1b. Switch S1, makes the selection. Fixed resistors R1 and R2 also form part of the timing chain. Resistor R2 is added to prevent the resistance of VR1 from going to zero at one end of the wiper travel. Capacitor C2 is decoupled merely to support the oscillator in the correct operating mode. The output waveform, a high-amplitude square wave, is available at pin #3. As it exits from IC1, the square wave is too harsh sounding for use in testing amplifiers and loudspeakers, since the inherent distortion of the square wave masks any amplifier distortions that are present. A sine wave would be ideal, but we can make an easy compromise with a triangular wave. Two components, R3 and C3, form a filtering network (an integrator) to reduce some of the signal harmonics by converting the square wave to a triangular wave. The resultant signal is thus more smooth sounding since the triangular wave is fairly close in wave shape to a sine wave. Note the "odd" value of resistor R3, which is 15 kohms. It is chosen to get the cleanest looking triangular waveform.

The amplitude of the converted signal will change as the frequency is varied, but this is a small discomfort when compared to the simplicity of the two-component convertor. The voltage amplitude is also attenuated by the convertor network, but this is again not an issue because only low-amplitude signals are needed for test purposes. Resistor R4 isolates the signal from the following stage. Capacitor C1 provides ac coupling and isolation into a final potentiometer. Since the absolute signal amplitude is dependent on the frequency, there is little point in calibrating the amplitude knob. The signal can thus be conveniently attenuated for testing pre-amplifiers or power amplifiers. Capacitor C5 provides the usual ac coupling into the external circuit. To finish off describing the oscillator, IC1's pin #1 is grounded, and pins #4 and #8 are taken to the positive supply voltage rail.

The frequency can be varied over the most useful part of the audio frequency band, about 64 Hz to 6 kHz. This frequency range is determined purely by the choice of the timing components used (R1, R2, VR1, C1a,b). The components selected come from a range that is most widely used, that is how we come up with this frequency band. The LM 555 timer IC can work down to a 5-volt supply, which is a useful extension of the operating life of the battery. The signal output level will decrease as the battery decreases, but this is of limited concern.

With the timing capacitor of 0.01 μF selected, the output frequency range spans 580 Hz to 6 kHz. With the timing capacitor of 0.1 μF selected, the output frequency range spans 64 Hz to 680 Hz. Since the capacitor ratio for C1a and C1b is 10:1, the frequency range varies by about the same ratio. The smaller the capacitor, the higher the frequency. The direct output from pin #3 is close to supply voltage amplitude. For testing purposes, use any speaker available and add a 1-kohm series resistor to increase the impedance (without the resistor, there will be severe loading on the output). You will hear a low-volume audio tone whose frequency will vary as the potentiometer VR1 is varied. The frequency should increase as VR1 is rotated clockwise. If this does not happen, reverse the connections to VR1. Toggle switch S1 to verify the change in frequency band.

Parts List

Semiconductor

IC1: LM 555 timer

Resistors

All resistors are 5 percent 1/4W
R1: 1 kohm
R2: 10 kohms
R3: 15 kohms
R4: 47 kohms
R5: 4.7 kohms

Capacitors

All non-polarized capacitors disc ceramic
All electrolytic capacitors 25V rating
C1a: 0.1 µF
C2b: 0.01 µF
C3: 0.1 µF
C4: 0.1 µF
C5: 0.1 µF
C6: 0.1 µF
C7: 100 µF

Additional Materials

VR1: 100-kohm potentiometer
VR2: 100-kohm potentiometer
D1: LED
S1: SPST miniature switch
S2: SPST miniature switch
Power supply: 9-volt battery

Construction Tips

Check out the functioning of the basic oscillator circuit first by using just a single capacitor for C1 (leave out the switch for the time being).

Potentiometer VR1 can be left out until later on, and the connection from pin #2 (of IC1) can be taken directly to R2. All the output components can also be left out, and you can temporarily hook up the output from pin #3 to a speaker (or headphones) via a capacitor with a 0.1 µF value (for example, C4). All the components associated with the power supply, which are located at the top of the power supply line, can be left out for now, too. With all this in mind, the circuit becomes very simple, and with so few components, the chances are very high that you will put together a working circuit on the first try. It doesn't matter what sort of a signal you get out of the speaker; if there's a tone, the circuit's fine. Now add the rest of the components in, knowing that the basic circuit works. This is a philosophy I always use when building a new circuit. It's so easy to make a connection to the wrong spot (the usual reason for circuit errors), and the more connections there are the more the chance for error. Use an IC socket for the IC because then it is very easy for replacements to be made (it's hardly likely that you'll need to, but just in case). Without an IC socket, removing an IC requires a lot of dexterity with the high risk of the board being damaged in the process. There are no low-level signals here, so wiring can be really any way you want it.

Test Setup

Once the circuit is fully assembled, you can do a full test. It's a good idea to feed the circuit into a speaker, because the high signal levels could cause unwanted distortion in headphones. In your testing you will check that

- volume level increases as the gain control VR2 is rotated clockwise
- frequency is increased when the smaller value of C1 is switched in
- frequency is increased as VR1 is rotated clockwise

In spite of the deceptive simplicity of the circuit, the output pseudo-sine wave is remarkably acceptable, and it certainly has nothing like the harshness from the indigenous 555's square wave. For our audio applications, this is more than adequate. There is no other integrated circuit that can compare with the 555 in terms of the ease of generating and controlling a signal source. You've got the best of both worlds with this circuit.

Project 10: Two-Input Mixer Pre-amplifier

An audio mixer such as this will allow you to overdub your voice onto a regular music track using just a handful of components in an elegantly simple 1-IC design. A regular mono cassette recorder (such as the piano-key type) is used as the music source, with the output fed straight from the ear-

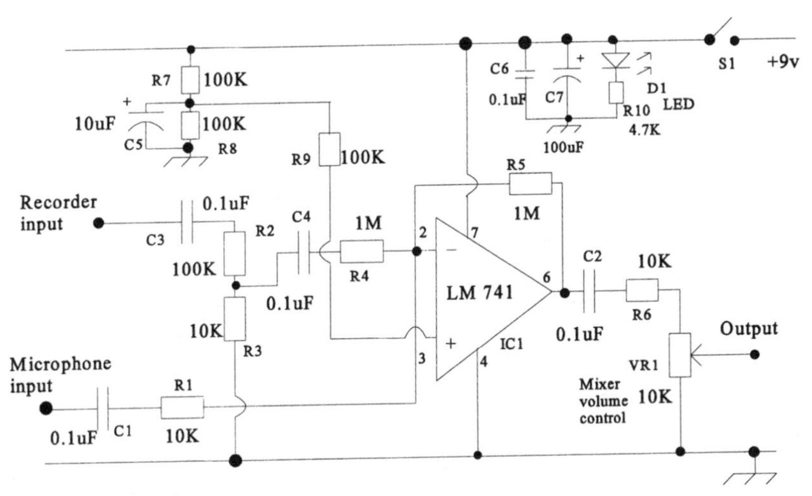

FIGURE 10-10 Audio mixer

piece socket. The output level will be high, of course, since the output has to drive a speaker, so it feeds into a simple two-resistor attenuator. The same carbon microphone that is supplied with the recorder is used to feed your voice onto the music track. A high gain stage provides the necessary boost for your voice. The LM 741 is configured in the inverting gain mode, but this time the two signals are combined in a basic summing mode. One of the nice features about using an op-amp is that novel circuits, such as this mixer, are so easily implemented. The cassette recorder's regular volume control is used to preset the music track to an appropriate level. Since there is no microphone level control, you control the music track instead. Simple voice overs can be quite impressive as you introduce a track while smoothly bringing up the volume. Fade-outs can be nicely done, too, as the music nears completion and you announce the next music selection. The mixer signal needs to be taken to a power amplifier so that you can listen to your DJ handiwork. The mixer volume control at the output stage gives you plenty of room to adjust for matching into your power amplifier. If it turns out that the microphone you're using is too sensitive, the stage gain can be easily reduced by reducing the value of the feedback resistor, R5. Make sure that R4 is also given the same value as R5. Singing along to your favorite tracks? Yes, that's something worth trying, too. It's amazing how much mileage you can get from a single integrated circuit. Some people might scoff at the thought of such a simple design and associate innovation with complexity.

One of the unique features in this book is the emphasis on novel designs using the bare minimum of active devices (integrated circuits). Some popular electronics hobby magazines have a characteristic trademark of publishing designs whose high component count is enough to restrict interest to the purist only. Notice how often the immensely popular LM 741 appears here. I've known many a professional designer to snigger at such a "low-tech" device when there are much higher performance op-amps on the market. But this collection is aimed at hobbyists, and, besides, exotic hard-to-get IC's would stop you dead even before you got started.

Circuit Description

So far, many of the earlier circuit descriptions have centered on the fundamental building block designs of the LM 741. You can probably find many similar projects in other sources, but this mixer circuit is totally original. It highlights the versatility that can be configured around just a single IC. Starting with the IC1 itself, we can see that there is a feedback resistor, R5, a 1-Mohm high-value resistor going from the output to the inverting input. So there has to be an accompanying input resistor. As it turns out there are two input resistors, R4 and R1, since this is a two-input mixer circuit.

Let's look at the simpler half first. The microphone input enters via capacitor C1, which in turn is connected in series to input resistor R1, which has a value of 10 kohms. From the knowledge we have gathered so far on the inverting amplifier we know that the gain is given by the ratio of the feedback resistor to the input resistor, that is, R5/R1 = 1 Mohm/10 kohms = 100.

Microphones only produce an output of a few millivolts, so the value after amplification would be a few 100 millivolts, which would match up nicely with a power amplifier of gain ×20 to give a final output signal of a few volts. The numbers all match up showing that the gain value of ×100 is about right. Note that the microphone for this circuit has to be one that does not require a dc bias voltage, for example, a simple carbon microphone. Other types, like the electret microphone, that require a bias will not function here. So before tearing your hair out wondering if this circuit works when it's built, remember the microphone restriction.

The non-inverting terminal is taken to the usual split bias voltage that is generated by two resistors of equal value, R7 and R8. Electrolytic capacitor C5 decouples the ac signal to ground. Isolation is provided by resistor R9. Do not leave this component out: it is necessary for correct circuit operation. The output signal is coupled via capacitor C2. A convenient volume control, VR1, allows for adjusting the signal level to suit the following stage. For further attenuation, resistor R6 is included. Even when VR1 is set to the maximum position (10 kohm), the inclusion of R6 (10 kohm) cuts the signal by 50 percent, since this would be merely two equal 10-kohm resistors in a simple fixed potential divider (see Figure 10-10).

Now it's time to go back to the second input. This half of the circuit is designed to accept a very high input signal level, that is, the output from a cassette recorder or portable radio. As we expect, there's a coupling capacitor, C3. The next two components, R2 and R3, form another potential divider attenuator. Because the incoming signal is much higher than the microphone signal, the attenuation is high, and the signal level is reduced by a factor of approximately 10. The junction between the two resistors is where the output signal is taken, coupled out, or connected, via capacitor C4 (0.1 µF). Finally, there's the matching input resistor (R4) for this circuit block. The value is 1 Mohm, so there's unity gain for this part of the mixer. That's all there is to it. As the circuits get progressively more complex, you can depend on the circuits being drawn and explained in a logical and consistent way to make understanding them very easy.

Parts List

Semiconductor

IC1: LM 741 op-amp

Resistors

All resistors are 5 percent 1/4W
R1: 10 kohms
R2: 100 kohms
R3: 10 kohms
R4: 1 Mohm
R5: 1 Mohm
R6: 10 kohms
R7: 100 kohms
R8: 100 kohms
R9: 100 kohms
R10: 4.7 kohms

Capacitors

All non-polarized capacitors disc ceramic
All electrolytic capacitors 25V rating
C1: 0.1 µF
C2: 0.1 µF
C3: 0.1 µF
C4: 0.1 µF
C5: 10 µF
C6: 0.1 µF
C7: 100 µF

Additional Materials

VR1: 10-kohm potentiometer
D1: LED
S1: SPST miniature switch
Power supply: 9-volt battery

Construction Tips

Start with just the microphone section—C1, R1, and R5 at the input—leaving out the recorder components for the time being (C3, R2, R3, C4, and R4). Don't forget to include the bias network going to terminal #3. By concentrating on just the bare essentials you can make sure the basic circuit is functioning. You will need a power amplifier of sorts to verify the output signal. Since a power amplifier is needed in any case to make use of this unit when it's completed, it's a good idea to pre-build one of the several power amplifier designs in this book, unless you already have one at hand.

Once you've verified that the circuit is working, add the balance of the components. Since you will need mechanical components, such as jack sockets, to facilitate the feeding of inputs and outputs, you might want to consider putting the circuit in a sturdy plastic project case. Use whatever sockets match your microphone and recorder outputs. The output socket can be matched to the power amplifier you decide to use.

If you're checking voltages and currents, the split bias point can be measured with a dc voltmeter. You should get a reading of half the supply voltage. Supply currents (measured in series with the positive supply voltage) are typically of the order of a few milliamps of current under quiescent conditions (when there are no signals passing through the stage).

Test Setup

When feeding this circuit into a power amplifier, make sure you switch on the power amplifier first, and then the mixer unit. The correct procedure is to start supplying power to the last circuit in the series connection and then the first circuit. Set the mixer volume control to midway, and send through your favorite music track, with the recorder volume set fairly low to begin with. Match your vocal input by adjusting the recorder volume control. The mixer volume control can be adjusted to suit the volume you want from your power amplifier.

Project 11: Pre-amplifier/Power Amplifier Combo

The pre-amplifier/power amplifier combo not only will be able to test a range of audio input devices (record sources, guitar inputs, microphones,

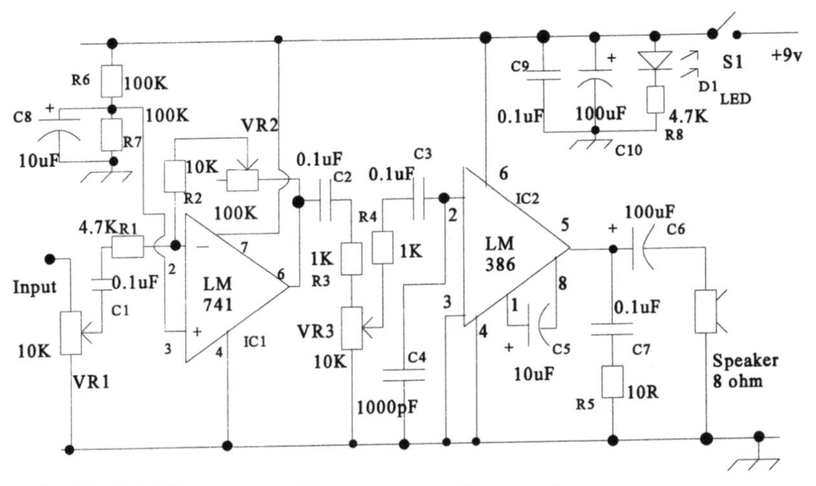

FIGURE 10-11 Pre-amplifier/power amplifier combo

turntables) but can also be used as a permanent dedicated audio amplifying unit. The generous range of adjustment of gain control made possible in this design (there are three gain controls here [VR1, VR2, VR3]) will give you the flexibility to cater to any type of signal input. The main volume control can be the final potentiometer feeding into the power amplifier once everything has been set up to your liking. If your preference is portability, then the speaker can be integrated into the project case, but bear in mind that the audio quality will be severely diminished. Better sound can be achieved using a decent-size speaker (at least 8 inches) mounted in its own speaker enclosure. If possible avoid putting the electronics inside the speaker cabinet. Although it might appear to be a good idea, it is better to house the electronics in a separate unit.

Circuit Description

This is the first twin IC circuit to be featured, so we'll examine it in sections. Let's start at the back end with IC2. This is a high-gain (×180) power amplifier using the LM 386 IC. We know it has a high gain because of the capacitor, C5, bridging terminals #1 and #8. There's a large, generous output capacitor, C6, emphasizing the low frequencies and feeding a speaker. A shunt Zobel network, consisting of capacitor C1 and resistor R5, takes care of maintaining a distortion-free output at high power levels. Because of the high gain coming in from the front-end stage (which we'll describe later), the signal is attenuated with resistor R4 and volume control VR3. The wiper terminal from VR3 feeds into a series resistor R4 and capacitor C3. The usual RF decoupling capacitor, C4, is connected between the input terminal (pin #2) and ground. Pin #3 is grounded. Pin #4 is the regular IC ground, and pin #6 runs to the supply voltage.

The first amplifier is a pre-amplifier based on an LM741, and it is run in the inverting mode with a gain of around ×23 (a total feedback resistance of 110 kohms divided by an input resistance of 4.7 kohms results in a gain of 23). A variable pre-amplifier gain is controlled by potentiometer VR2. Fixed resistor R2 limits the minimum feedback resistance to 10 kohms (resulting in a gain of 10 kohms/4.7 kohms = about ×2) when VR2 is set to it's minimum value. The ac coupling is provided by capacitor C1. With the three potentiometers in this circuit you have a lot of flexibility to examine the effects of each. Generally, VR1 will be matched to the input signal level, mainly attenuating higher level inputs. VR2 sets the pre-amplifier gain such that when the following potentiometer, VR3, is set to a maximum setting, the final output is still distortion free. Watch out for the polarity requirements of capacitors C3 and C5.

Parts List

Semiconductor

IC1: LM 741 op-amp
IC2: LM 386 audio power amplifier

Resistors

All resistors are 5 percent 1/4W
R1: 4.7 kohms
R2: 10 kohms
R3: 1 kohm
R4: 1 kohm
R5: 10 ohms
R6: 100 kohms
R7: 100 kohms
R8: 4.7 kohms

Capacitors

All non-polarized capacitors disc ceramic
All electrolytic capacitors 25V rating
C1: 0.1 µF
C2: 0.1 µF
C3: 0.1 µF
C4: 1000 pF
C5: 10 µF

C6: 100 μF
C7: 0.1 μF
C8: 10 μF
C9: 0.1 μF
C10: 100 μF

Additional Materials

VR1: 10-kohm potentiometer
VR2: 100-kohm potentiometer
VR3: 10-kohm potentiometer
D3: LED
S1: SPST miniature switch
8-ohm speaker
Power supply: 9-volt battery

Construction Tips

Keep the wiring connections short and neat since there is a healthy quantity of gain to be obtained from the two combined amplifiers: you've got about ×23 maximum pre-amplifier gain and ×180 with the power amplifier, giving a total gain of over ×4000. Short wiring helps keep the circuit from oscillating and becoming unstable. Also keep the input connections away from the output connections.

It's a good idea to build the combo in stages and verify the first section before moving on to the next. In that way the troubleshooting of the circuit will be much less tedious if the circuit doesn't work at switch-on. Components like C4 (the RF bypass capacitor), C5 (the high-gain-setting capacitor), C7, and R5 (the Zobel network), can be temporarily left out until the circuit has been checked out, since leaving out these components doesn't affect the basic operation of the circuit. Where there's more than one IC in the circuit I always build one section first, verify that it works, and then move onto the next. It doesn't matter which half you start with. Personally I find the power amplifier cannot go wrong, because it doesn't need the split bias voltage (as does the LM 741), so that makes it less likely to go wrong. As it turns out, the LM 386 can be powered up with just three basic components—the IC itself, the input capacitor, and the output capacitor—so you can quickly do a check on whether your connections are OK.

Test Setup

The circuit test setup is simplicity itself. First set VR1 to the midpoint position. Any input signal can be connected into the combo because VR1 takes care of any amplitude variations. Then set VR2 to the 0 ohm position, so the gain of the first stage is at a minimum of ×2 (VR2 should be rotated in the counterclockwise direction for the minimum gain position). Should you find that you don't get a minimum gain, swap the outer terminal connection of VR2 to the other outer terminal. Finally, set VR3 to a midpoint position.

Depending on the level of the signal input you could get either a weak, a strong, or perhaps an overloaded signal. If it is too loud, reduce the signal input at the source (turn down the radio, if it is a radio signal), or reduce VR1 if the input is a fixed signal. You can then play around with increasing the stage gain of IC1 by gradually increasing VR2. The sound level will increase dramatically. Then vary VR3 to confirm that you have full control over the signal output level.

The interaction between the three volume controls is quite interesting to explore. Each volume control serves its own purpose. Start with the final control, VR3. When VR1 is set to the maximum volume position and the signal being fed to it is kept to a level that produces no distortion, you can keep VR1 and VR2 fixed but vary VR3 over its full range to produce a signal from zero to full amplitude. Now consider the first volume control, VR1. If the input signal were to come from widely different sources, with different amplitudes, then VR1 is used to make sure that when the final volume control, VR3, is turned up to its maximum value there is still a clean signal coming out. Finally, when the input signal is weak and VR1 is already turned up to its maximum value, you can now get more gain by increasing the value of VR2.

Project 12: Microphone Test Set

This test set provides a lot more analytical capability than just verifying that a microphone works. Any other form of transducer that converts sound energy into electrical energy, such as piezo-electric elements, can be measured using this test set. This is a novel two-IC circuit, but actually it's a single package, since both the amplifiers are in the same package. To be more specific, there are actually four op-amps in the same component, but only two are needed here. A regular buffer (you've come across this circuit earlier) is added to provide the isolation needed to drive the low-impedance analog meter indicator; without it, the circuit couldn't function. And in between the two ICs there's a novel little diode circuit that changes an ac signal into a slow-moving dc signal. That's the signal that drives the meter ultimately.

Right at the front end a useful switched bias arrangement nicely accommodates either type of microphone, the biased and unbiased type, so there's no need to search around and figure out whether or not a microphone needs a bias. Plug in the microphone and simply tap on the face of it; if the meter shows no sign of a deflection, switch the bias. It'll work with one of the two options.

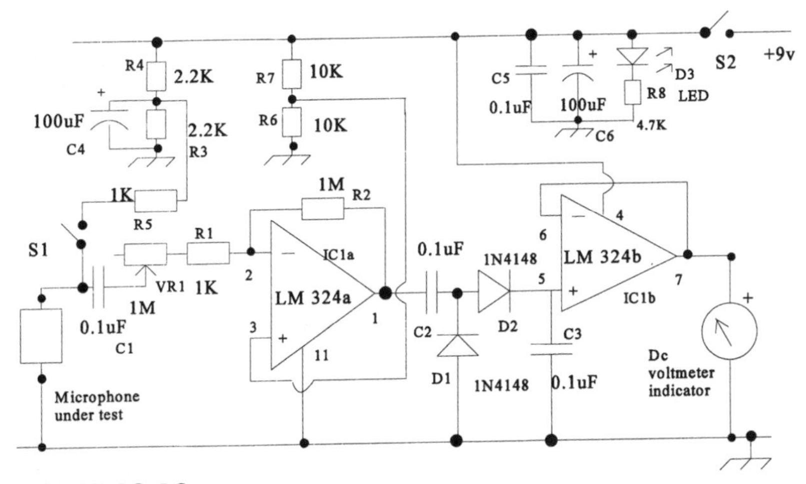

FIGURE 10-12 Microphone test set

Just to show that the gain variation for an op-amp can also be achieved by varying the input resistor (VR1 and R1 are the combined input resistor network), rather than the more normal feedback resistor (R2), that's the arrangement that is shown here. And the customary shunt capacitor across the midpoint of the split supply bias voltage is also eliminated. This application is not really "audio" in the true sense, so this arrangement is fine.

Circuit Description

Microphones convert sound energy into electrical energy. This test set provides a useful means of checking if a microphone works, and it also helps you determine the relative sensitivity of different types of microphones. In terms of their operational modes, microphones commonly fall into two categories: either they require no power, as in the case of simple carbon microphones; or they require a small dc bias voltage, as in the case of newer electret microphones. If you have a collection of microphones or microphone inserts, this test set will also help you sort them out.

There are three distinct parts to this circuit. The most unusual is the indicating device shown at the output of IC1b. This is a regular analog dc voltmeter or, more specifically, your multimeter set to the dc voltage range. The positive (red terminal) goes to the IC's output pin. You need to see the response of the needle, so don't substitute for the analog type. A digital meter will work, but a flickering sequence of digital numbers will be almost meaningless. Besides, the analog meter doesn't require a battery to operate (unlike it's digital cousin).

The other unusual item you'll notice, if you've been following the circuit projects in sequence, is a new IC, the LM 324. It is a quad op-amp; that has four separate ICs in one component. This 14-pin IC's connections are totally different from those of the popular LM 741. So be very careful with the wiring sequence if you haven't used this kind of IC before. This multi-IC device takes up less space, and it is commonly featured in audio amplifiers and filter circuits.

Starting at the front end of the circuit we see the microphone under test input. Two terminals are required for the input, so use whatever convenient connector you prefer. One option is a regular miniature jack socket, but that means you've got to have the mating jack plug on the microphone. Or you can just have separate screw-type or push-type terminals. It really doesn't matter. The first integrated circuit, IC1a (the *a* indicates it's part of the same IC), is configured as a regular inverting mode amplifier. Note that the IC pin numbers are totally different from those for the LM 741, and there's only a ground connection on IC1a (the power connection is on IC1b). Feedback resistor R2 is high (1 Mohm), because we will be requiring a lot of gain (microphones don't provide a lot of signal). The input resistance is a combination of fixed resistor R1 (1 kohm) and potentiometer VR1 (1 Mohm). When VR1 is set at a minimum, the total input resistance is just 1 kohm and the gain is thus 1M ohm/1 kohm = ×1000. Input capacitor C1 couples the microphone input to VR1's wiper terminal.

As it stands, carbon microphones requiring no dc bias would work with this part of the circuit. The additional circuitry used to power electret microphone comes from resistors R3 and R4 and capacitor C4. Resistors R3 and R4 are of equal value; hence the bias voltage is half the supply voltage. *Note:* The actual 2.2-kohm values are lower than the usual 100-kohm values we've been using for biasing the op-amp. In this instance the component values are dependent on the circuit position they're used in. An isolating resistor, R5, with a fairly low value of 1 kohm goes to switch S1. The switch allows you to select the bias at will, thereby providing versatility. The bias, as we've mentioned before, is needed for electret microphones to operate properly. Returning to IC1a, a split supply bias voltage is needed for the positive terminal (pin #3). This is derived from a network of two resistors of equal value, R6 and R7, each equal to 10 kohms.

The signal leaving the first IC through capacitor C2 is an ac signal. We need to convert this to a dc signal in order to drive the indicating device (the voltmeter). Small-signal silicon diodes D1 and D2 and capacitor C3 form a rectifying network such that the output across C3 moves in sync with the original microphone signal. Capacitor C3's value is chosen so that the changing dc voltage is maximized and at the same time the response of the rectifying circuit moves in step with the input signal. The 0.1 μF value is a good choice. If you were to connect a voltmeter directly across capacitor C3, there would be no output, since the voltmeter would significantly shunt the signal. A buffer stage is what is needed. The buffer has a high input impedance, so there's no loading on the rectified signal, and the buffer has a low output impedance, so the voltmeter doesn't load it (the buffer). There's a direct connection between output and input (in this case the negative input). The signal goes directly to pin #3. Because this is a dc amplifier, there is no usual split supply biasing arrangement. *Note:* As the rectified signal input goes to the non-inverting terminal, the buffer output is in phase with the input; that is, if the input rises, the output rises, and vice versa.

Parts List

Semiconductor

IC1a: 1/4 LM 324 quad op-amp
IC1b: 1/4 LM 324 quad op-amp

Resistors

All resistors are 5 percent 1/4W
R1: 1 kohm
R2: 1 Mohm
R3: 2.2 kohms
R4: 2.2 kohms
R5: 1 kohm
R6: 10 kohms
R7: 10 kohms
R8: 4.7 kohms

Capacitors

All non-polarized capacitors disc ceramic
All electrolytic capacitors 25V rating
C1: 0.1 μF
C2: 0.1 μF
C3: 0.1 μF
C4: 100 μF
C5: 0.1 μF
C6: 100 μF

Additional Materials

VR1: 1-Mohm potentiometer
D1: Silicon signal diode 1N4148

D2: silicon signal diode 1N4148
D3: LED
S1: SPST miniature switch
S2: SPST miniature switch
Power supply: 9-volt battery

Construction Tips

The circuit can be comfortably tested in two parts. Take special care with connecting up the power and ground terminals for the IC. Start with the construction of the front end, the microphone pre-amplifier, up to the point of capacitor C2. A suitable test microphone would be the type supplied with the piano-key type of mono cassette player. These usually have a 1/8-inch mono jack plug at the end of the microphone cable. You can verify that the microphone is working by recording something. These old-style microphones typically do not require a bias voltage to operate, so switch S1 should be in the off position. Verify S1's toggle positions with a continuity tester or an ohmmeter and mark the on/off positions. Feed the output from C2 into a power amplifier to verify that the circuit is functioning. VR1 will control the gain. Check to see that the gain increases with a clockwise rotation of VR1. If this is not the case, swap the connection to the other outer terminal. If you have an electret microphone, check this out with the bias connection through S1. Electret microphone inserts are commonly available from electronics component stores and are considerably smaller than the older carbon microphone inserts.

Once you have verified the first section of the circuit, go on to test the second section. Temporarily feed an ac signal through C2, which must be disconnected from the output of IC1a. Couple an analog dc voltmeter across IC1b's output terminals. As the amplitude of the input voltage is increased, the voltmeter needle will be deflected further. Make sure the diode polarities are positioned as shown in Figure 10-12.

Test Setup

Principally this test set will verify any microphone for you. Simply plug in the microphone and check the deflection on the meter when you tap the face of the microphone.

If you have a collection of microphones, there's a much more interesting experiment you can carry out. You can determine the relative sensitivities of various microphones by using a constant sound source, such as an audio test generator feeding into a power amplifier and speaker. Set up the sound source a fixed distance from each microphone and adjust the gain control for a midpoint deflection on the voltmeter monitor. Record the results in a table so you have a gauge of the relative sensitivities of all the microphones in your collection.

This test set can also be used to measure the directional response of a microphone. Mount the test microphone on a platform that can swivel in the horizontal plane. Vary the angle of the microphone (use a simple protractor to measure the angle) and read off the corresponding response.

Project 13: Audio Test Set

If you've just built a project, chances are you're going to need to test it. A signal source and a means of monitoring signals are the most essential and versatile pieces of test equipment you can have, next to your trustworthy multimeter. Invariably first-build circuits do not work the first time we switch them on, and a little troubleshooting with the audio test set will put you on the path to a working circuit.

The audio test circuit is a combined signal generator, power amplifier, and self-contained speaker. Together these components provide the versatility you need to feed various signals into and out of the test set. Practically any test option you want is available through the test sockets and switches designed into the circuit. Two ICs carry the workload, generating the signal source and providing the capability for driving a low-impedance speaker load. Testing headphones and speakers, checking power amplifiers, evaluating pre-amplifiers, even doing continuity checks—that's a lot of functionality in one circuit.

Circuit Description

Two separate circuits are combined here into a versatile audio test set. The square-wave generator based on the LM 555 is connected into an LM 386 power amplifier to form the basis of a versatile diagnostic circuit for

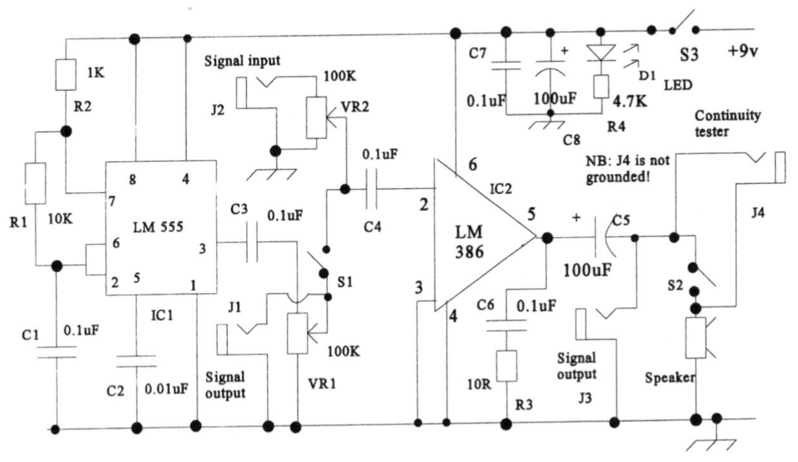

FIGURE 10-13 Audio test set

testing speakers, pre-amplifiers, microphones, and almost any variety of audio device. The versatility stems from the inclusion of a number of strategically placed switches and jack sockets. A continuity tester for evaluating cable runs is also provided in this test set.

IC1 is a fixed-frequency LM 555 free running square-wave oscillator generating a convenient audio frequency (i.e., you can hear it). Resistors R1 and R2, together with capacitor C1, form the timing network. The frequency generated doesn't matter provided it's located somewhere convenient in the audio frequency band. Capacitor C2 sets up the correct operation for IC1. The ground connection is made to pin #1. Power is supplied to pins #4 and #8. The output signal exits through capacitor C3. A potentiometer, VR1, provides continuous reduction of this high-level signal to a more useful level. Jack socket J1 is a parallel take-off point for the feed from VR1's wiper. This variable level feed can be used to check out speakers. The signal will be attenuated by the low speaker impedance, but for a quick go/no-go test, this is great. If you have an oscilloscope, here's a useful waveform test source.

The second stage is a fairly conventional audio power amplifier using the LM 386. The output feed is to a small speaker, perhaps an inch or two in diameter, since the speaker's only purpose is to signify the presence of a signal that is successfully passing through. The input feed to IC2 is to pin #2 via capacitor C4. There is a switch feeding the signal from VR1's wiper terminal to C4. The reason for the switch is to isolate the LM 555 signal, so an external source can be used. This route is taken care of by the second jack socket, J2, and potentiometer VR2. When switch S1 is in the off position, an external signal can be fed into the power amplifier. So if you've built the signal generator, "Project 9: Audio Signal Generator," you can check the signal generator out by running it though the power amplifier stage. Since there will be large amplitude signals emerging from the output of IC2, a Zobel network (C6 and R3) is included, so that the high-level output will be distortion free. Electrolytic capacitor C5 feeds the power amplifier signal to the speaker load. Another jack socket, J3, couples this output for use as a feed into an external speaker for those times when you want to make full use of the amplified signal output. An 8-inch speaker properly mounted in an enclosure will produce considerable listening fidelity and volume. The small internal speaker can be muted with switch S2. Finally, there's jack socket J4 bridging switch S2. Jack socket J4 will serve as a very useful continuity tester by plugging in a mating jack plug that has two flexible test leads.

Parts List

Semiconductor

IC1: LM 555 timer
IC2: LM 386 audio power amplifier

Resistors

All resistors are 5 percent 1/4W
R1: 10 kohms
R2: 1 kohm
R3: 10 ohms
R4: 4.7 kohms

Capacitors

All non-polarized capacitors disc ceramic
All electrolytic capacitors 25V rating
C1: 0.1 µF
C2: 0.01 µF
C3: 0.1 µF
C4: 0.1 µF
C5: 100 µF
C6: 0.1 µF
C7: 0.1 µF
C8: 100 µF

Additional Materials

VR1: 100-kohm potentiometer
D1: LED
S1: SPST miniature switch
S2: SPST miniature switch
S3: SPST miniature switch
J1: 1/8-inch mono jack socket
J2: 1/8-inch mono jack socket
J3: 1/8-inch mono jack socket
J4: 1/8-inch mono jack socket
Power supply: 9-volt battery

Construction Tips

Leave the power supply capacitors and LED circuit, jack sockets, and switches until the very end. Start by verifying that the oscillator is working. The output from IC1 can be checked temporarily by feeding the output through C3 to an external speaker via a couple of test leads. Since this audio test set can get a lot of use, a project case is definitely needed. Once the circuitry around IC1 has been verified, move on to the building of the power amplifier. As a preliminary test you need only the input coupling capacitor,

C4, and the output capacitor, C5. Since the oscillator has already been verified, the power amplifier verification is very simple. Connect up the external speaker to the amplifier's output and route the oscillator signal into capacitor C4. The signal output will be extremely loud given the magnitude of the oscillator signal—but only a momentary contact to C4 is needed. If further testing is needed, you can wire in potentiometer VR1 and use a more manageable test signal.

Test Setup

1. *Continuity tester.* S1/on, S2/on. Adjust VR1 for the required signal level (a comfortable listening level) from the internal speaker. Open S2. Plug test probe into J4. Use test probe leads as continuity tester. A tone will sound when there is an electrical connection across the probes.
2. *External speaker check.* S1/on, S2/on. Adjust VR1 for required signal level from internal speaker. Plug in jack plug to J3. Connect jack plug leads to external speaker. Open S2. Signal will be passed to external speaker. If external speaker is not defective, signal will be heard.
3. *External power amplifier check.* S1/on, S2/on. Adjust VR1 for required signal level from internal speaker. Open S1. Plug in jack plug to J1 and feed signal to external power amplifier with known good speaker. Good test signal output indicates that the power amplifier is OK.
4. *External low-level signal check.* S1/off, S2/on. Feed external signal through J2. Adjust VR2 for required signal level from internal speaker. Audio output indicates good signal source.

Project 14: Guitar Pre-amp and Buffer

A little extra bite can make all the difference to a lead guitar's signal especially if you're running extra-long cables to your amplifier. With all of the band's input added to your lead guitar work, you might want to just record your lead track. This project is a two-IC circuit providing pre-amplifier bite for your Stratocaster plus a nice buffer for isolating the final feed. This circuit has a neat recorder take-off point for feeding the pre-amplifier into any stock piano-key type mono recorder. Although the piano-key-type cassette player might seem a little archaic compared to the modern Walkman-style stereo cassette player, it's nevertheless very robust, inexpensive, and has a convenient mono input and mono output.

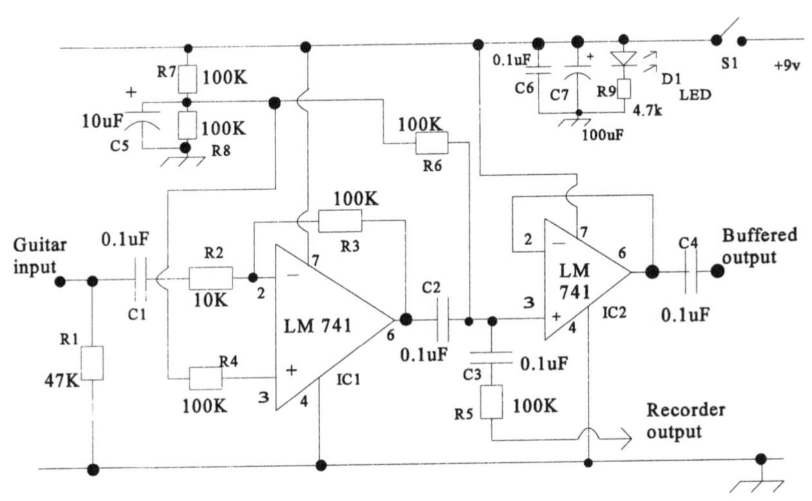

FIGURE 10-14 Guitar pre-amplifier and buffer

Final processing into the main guitar power amplifier is taken from the buffer's output. The circuit is nice, clean, and simple—there's nothing critical about the design or construction. Although the inclusion of the buffer might seem superfluous, don't omit it; it really serves a purpose in spite of its deceptive simplicity. Guitar signals are not noted for quality, so the LM 741s do more than adequate justice to the guitar's signal-processing requirements. This circuit is good for generating a signal feed into a recorder. Normally with a feed into a standard guitar amp, a recorder take-off point is not readily accessible. When the feed is taken from the amp's headphone socket, the signal level is too high, and also the main speaker gets shut off. As an amateur Strat blues player, I've used this circuit many times to check out my fretboard skills, and personally I find it very useful, so for all you budding guitarists, I've included it here. In fact some of the other circuits will also testify to my preferences in the guitar field.

Circuit Description

Two LM 741s are used in this unusual design. Starting at the front end, an electric guitar input feeds in through capacitor C1. Most likely the coupling will be made through a 1/4-inch jack socket. A load-matching resistor of 47 kohms (R1), commonly found in guitar amplifiers, is connected directly across the input. IC1 is set up in the inverting mode to produce a gain of ×10 via the feedback resistor R3 (100 kohms) and the input resistor R2 (10 kohms). The IC's positive input, as we would expect, requires a split voltage bias supply, and this is supplied through a resistor combination of R7 and R8, resistors of equal value connected directly across the supply voltage. A capacitor, C5 (10 µF) shunts the ac component and provides a direct link to ground. Resistor R4 provides a degree of isolation, because, as we will see later, the next IC also shares the same split supply bias. IC2 is a buffer, and the output terminal is connected back to the negative input terminal as a requirement of this configuration. The input signal is fed to the positive input terminal via capacitor C2. At the same time, there has to be a split supply bias brought to the same terminal. This is done by taking a resistor, R6, to the same previous split supply bias network, that is, the junction of R7 and R8. The two resistors R6 and R4 are thus isolating resistors.

An unusual additional circuitry is added to pin #3 of IC2. This take-off feed, connected via capacitor C3 and resistor R5, can be fed to a recorder for taping your guitar signal. Finally the buffered output is taken from capacitor C4. This can be fed into a regular guitar amplifier or a power amplifier.

Note: Electric guitars generate a huge amount of harmonics especially when chords are played. A heavy-duty speaker designed specifically to take theses harsh signals is essential. A regular speaker is very likely to be destroyed under continual use.

Parts List

Semiconductor

IC1: LM 741 op-amp
IC2: LM 741 op-amp

Resistors

All resistors are 5 percent 1/4W
R1: 47 kohms
R2: 10 kohms
R3: 100 kohms
R4: 100 kohms
R5: 100 kohms
R6: 100 kohms
R7: 100 kohms
R8: 100 kohms
R9: 4.7 kohms

Capacitors

All non-polarized capacitors disc ceramic
All electrolytic capacitors 25V rating

C1: 0.1 µF
C2: 0.1 µF
C3: 0.1 µF
C4: 0.1 µF
C5: 10 µF
C6: 0.1 µF
C7: 100 µF

Additional Materials

D1: LED
S1: SPST miniature switch
Power supply: 9-volt battery

Construction Tips

Start by building the pre-amplifier without the buffer. That way you cut down on the difficulty of troubleshooting if the circuit doesn't work at first switch-on. You'll also need a power amplifier or guitar amplifier to check out this stage of the project. Apart from the extra input resistor, R1, the circuit is identical to the rest of the inverting pre-amplifiers that figure in this book, so if you've already started with the earlier circuits, this'll be a cinch. You should get a gain of ×10 by going through the pre-amplifier. There will be a significant difference when you compare the guitar signal fed directly to the power amplifier with that boosted by the pre-amplifier. Once that stage is correct, go on to add the following buffer stage. The signal level put through the buffer will come out the same, as if the buffer weren't there, since the buffer is a unity gain device. But it's not the gain that we want; it's the buffer's isolating function.

Test Setup

You need to feed the buffer into a guitar power amplifier. A regular mono recorder's input is coupled to R5. Use whatever socket is most convenient, probably a 1/8-inch jack socket/plug combination would be a good choice. Any electric guitar (Stratocaster) can be used for the input signal. Adjust the guitar's volume control to suit for a distortion-free signal into the recorder output, since different pickups produce different output levels. The recorder output will be a faithful rendition of your riffs. Once more, the use of the innocent looking buffer is demonstrated. Having seen the buffer's usefulness in more than one application by now, you can probably see that for a circuit that on the surface appears to do very little, this one has a very special role to play.

Project 15: Guitar Fuzz Pre-amplifier

For any electric guitarist who is also into electronics, the simple fuzz circuit is a really great way to check out a cool-sounding effect. Strange as it may seem to hi-fi enthusiasts, fuzz or distortion can be used as a guitar effect. A simple way to get distortion is to crank up your power amplifier to the upper limit and wait for your speakers to blow themselves into oblivion, like the legendary Jimi Hendrix running his Marshall stacks into a mind-blowing and ear-splitting cauldron of fiery pentatonics. But emulating your heavy metal guitar icon may not be a good way to ingratiate yourself with your neighbors. A much more sophisticated technique is to use a very high gain pre-amplifier that is driven well into distortion (sine waves will start to clip and look like square waves when the distortion happens). When you follow the distorted pre-amplifier signal with an even small power amplifier like the LM 386 (less than a watt of power), you still get a terrific fuzz sounding effect. Incidentally, fuzz is also synonymous with the term *overdrive*. Start with a highly overdriven first stage, reduce the signal level, and follow it with a power amplifier (not necessarily high power, because the distortion has already taken place). With distortion comes inherent added sustain as a result of the overdriven clipped waveform. The Rolling Stone's classic opening intro bars to "Satisfaction" is the all-time most recognizable example of fuzz—probably one of the earliest mid-1960s recordings that makes use of this effect. Fuzz is distortion or overdrive; it gives guitar notes and chords that raspy sound. There's some sustain also because the overdriven notes take a longer time to die away than the clean undistorted sound. Incidentally, with a little fuzz thrown in for good effect, it's deceptively easy to sound better than you really are.

Circuit Description

With a little bit of ingenuity even a single IC design can produce a remarkable circuit. This one uses the LM 741 to produce a switch-selectable guitar pre-amplifier with fuzz option thrown in.

IC1 is configured as an inverting amplifier with a feedback resistance of VR1, a 1-Mohm potentiometer, and R3, a fixed-value resistor. The input resistor is R3, a 2.2-kohm component. When VR1 is at its minimum position, the gain is unity; at the high setting of VR1, the gain peaks around ×455, a very high gain value. Under normal conditions this setting isn't used because of the distortion produced. Normal gains of around ×10 are what are generally used. The input capacitor C2 has a fairly high value so that it will pass the lower bass frequencies through. The usual impedance-matching 47-kohm resistor, R1, shunts the guitar signal input. A small capacitor, C1, pre-

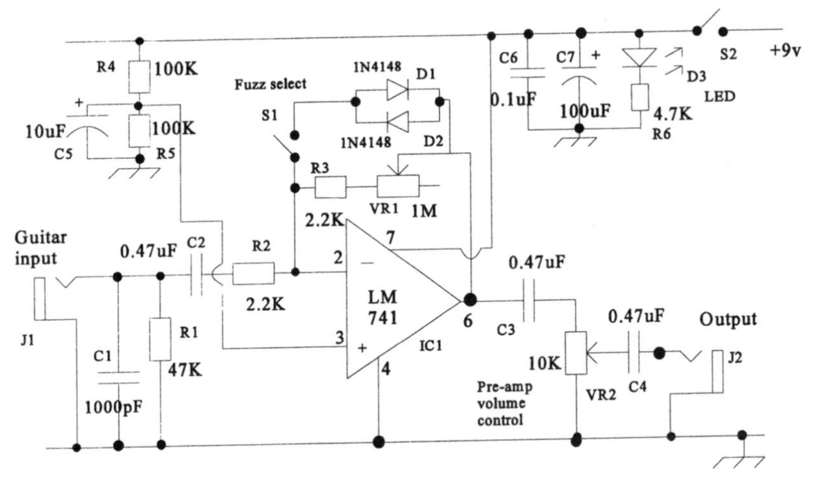

FIGURE 10-15 Guitar fuzz pre-amplifier

vents any radio frequency interference from being picked up. At the high-gain value, picking up a radio frequency will certainly be a possibility. Guitar input jack socket (1/4 inch) is J1. There are two other components that shunt the feedback resistance. Silicon diodes D1 and D2 are wired back to back and provide the nice clipping function that produces the fuzz effect. Switch S1 selects either normal or fuzz mode. In the fuzz mode the gain has to be set to its maximum value. The output signal exits through capacitor C3, and since the volume in the fuzz setting is very high, a potentiometer, VR2, is added to attenuate the signal to a useful level. The regular wiper output is taken through a capacitor, C4, into output jack socket J2. Bias for the non-inverting terminal is generated through the equal resistor network of R4 and R5. Capacitor C5 provides an ac ground path.

Parts List

Semiconductor

IC1: LM 741 op-amp

Resistors

All resistors are 5 percent 1/4W
R1: 47 kohms
R2: 2.2 kohms
R3: 2.2 kohms
R4: 100 kohms
R5: 100 kohms
R6: 4.7 kohms

Capacitors

All non-polarized capacitors disc ceramic
All electrolytic capacitors 25V rating
C1: 1000 pF
C2: 0.47 µF
C3: 0.47 µF
C4: 0.47 µF
C5: 10 µF
C6: 0.1 µF
C7: 100 µF

Additional Materials

D1: LED
VR1: 100 kohm potentiometer
VR2: 100K potentiometer J1: 1/4-inch jack socket
J2: 1/4-inch jack socket
S1: SPST miniature switch
S2: SPST miniature switch
Power supply: 9-volt battery

Construction Tips

Begin the build with just the basic pre-amplifier components. Leave out the shunt diodes, D1 and D2, until the end—until the amplifier is verified. Some other secondary components are C1, R1, VR2 and C4, which can be omitted until later, since their absence will not affect the operation of the circuit. VR1 is needed to control the pre-amp gain. At the maximum position of VR1, there could be distortion occurring because of the high stage gain setting.

Test Setup

Couple a Strat (or whatever you have handy) into the input and take the output into any guitar power amplifier. Make sure the speaker is a heavy-duty instrument unit that can handle heavy distortion, if you want the speaker to last. Adjust to maximum gain and keep VR2 low. The main power amplifier can be kept at a low volume. Notice the distinct differences when you switch S1 to the clean position (the gain setting might require changing). A straight guitar signal is clean and has relatively little sustain. Check out the sound difference for a chord riff. The power amp's output can also be fed into headphones if you wish.

Project 16: Electric Guitar Pacer

You've seen them before—guitar wannabes playing air guitar. If you're a budding guitar player, here's a chance to go one better. The superb realism of this inexpensive, easy-to-build audio guitar project will blow you away. All budding electric guitarists want to emulate their heroes, be it the blues greats (Eric Clapton, B.B. King, Buddy Guy, Robert Cray, and so on) or country's great guitar pickers (Chet Atkins, Vince Gill, Lee Roy Parnell, and so forth). The electric guitar pacer allows you to seriously play/track alongside the master and copy his or her riffs, note for note. Searing pentatonics, hammer-ons, pull-offs—trade them all with B.B's "Lucille." You can't get any closer to your favorite performer than with the combined signals from your guitar and your cassette player, coming through headphones. When you match the sound of your guitar against the soaring riffs of guitarland's greatest using your electric guitar pacer, you'll swear you're live in the studio with Jimi Hendrix. Trade searing licks with bluesmaster Eric Clapton's Stratocaster as he takes Leo Fender's stunningly inspirational creation through its paces. Or you can dreamily provide backup for Nashville's gorgeous Lorrie Morgan as she pours her heart out for you.

Since there are so few parts in this project, you can be up and running within an evening and still have time to start jamming away into the night. Dual-channel volume controls enable the optimum flexibility match between you and the "band." The combined tracks playing back through headphones oozes with realism.

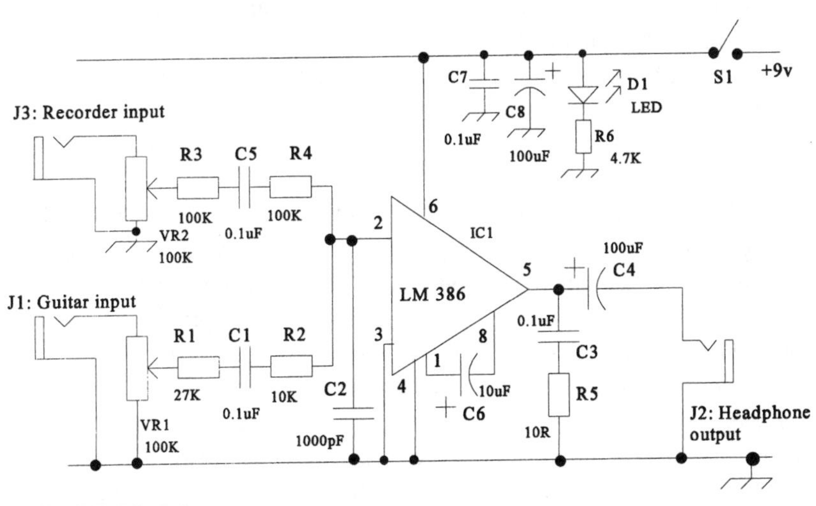

FIGURE 10-16 Electric guitar pacer

Circuit Description

The heart of the electric guitar pacer is the LM 386 low-power audio amplifier IC, running off a 9-volt battery. Independent volume controls are provided for the electric guitar input and the cassette recorder input. In this way you can nicely adjust the relative signal volumes for both sources without scrambling around the cassette recorder controls and the guitar controls. What does the electric guitar pacer sound like? Even with a regular mono cassette recorder and budget quality headphones, the combined sound is terrific: it really sounds like you're one of the band members!

The feed input for the guitar is through a 1/4-inch jack socket, and a 1/8-inch miniature jack socket is used for the recorder input and headphone output. The trick lies in having both feeds come through headphones rather than a speaker—extraneous room sounds are blocked out, and the effect is startlingly impressive.

As we've said before, IC1 is the ever popular LM 386. Since there is very little power being used, the lower-power LM 386 N-1 variant is specified here, but other variants can also be used if you've got them on hand. As configured here, the IC has a gain of ×200, which is good enough for our application. There is no point in having any more gain, since all you'll be doing is deafening your ears. You don't need a lot of power to get devastatingly loud headphone volumes.

The guitar input feeds into J1, a standard 1/4-inch jack socket, coupled to potentiometer VR1 for volume control. The feed, from VR1's center wiper terminal, is capacitively routed through resistor R1, capacitor C1, and resistor R2. Resistors R1 and R2 are needed for an optimized isolation between the guitar's volume control and the recorder input, so there is minimum loading by the guitar's controls. Capacitor C1 is the usual ac coupling capacitor. R2 is connected to pin #2 of IC1, and at the same time it is also connected to a small shunt capacitor, C2, to prevent RF breakthrough at the input. A conventional Zobel shunt network, made up of resistor R5 and capacitor C3, shunts the output from pin #5, stabilizing the sound output at higher levels. Capacitor C4 finally takes the output to headphone jack socket J2.

Going now to the recorder input, we see that the signal feed comes in through jack socket J3. Potentiometer VR2 controls the volume of your feed into resistor R3, capacitor C5, and resistor R4. Capacitor C5 provides the usual ac coupling. Resistors R3 and R4 are for isolation, preventing the guitar input feed at pin #2 from being shunted to ground. At the power end of the circuit, capacitors C7 and C8 provide decoupling for IC1. Finally LED D1 and current-limiter resistor R6 provide a useful means for a power "on" indicator. If you want, you can leave them out, but I always like to know when the power is on. It saves on having to replace run-down batteries. To obtain the gain of ×200, capacitor C6 is needed to bypass pins #1 and #8.

Normally the LM 386 has a low gain of only ×20 without C6. IC1's ground pins are #3 and #4. S1 is the power-on switch. As you can see in Figure 10-16, the circuit is very simple, but the design smarts incorporated minimize mutual signal loading between the two sources.

Parts List

Semiconductor

IC1: LM 386 audio power amp

Resistors

All resistors are 5 percent 1/4W
R1: 27 kohms
R2: 10 kohms
R3: 100 kohms
R4: 100 kohms
R5: 10 ohms
R6: 4.7 kohms

Capacitors

All non-polarized capacitors disc ceramic
All electrolytic capacitors 25V rating
C1: 0.1 µF
C2: 1000 pF
C3: 0.1 µF
C4: 100 µF
C5: 0.1 µF
C6: 0.1 µF
C7: 0.1 µF
C8: 100 µF

Additional Materials

D1: LED
VR1: 100K potentiometer
VR2: 100K potentiometer
J1: 1/4-inch jack socket
J2: 1/8-inch mono jack socket
J3: 1/8-inch mono jack socket
S1: SPST miniature switch
Power supply: 9-volt battery

Construction Tips

Check out the main power amp first. You can cut it down to just the basic components first. Use just the guitar input components, leaving out the Zobel network (R5, C3) and gain capacitor (C6). The C2 capacitor can also be dropped initially. With the input socket (J1) and output socket (J2) in place, especially if they're mounted in a sturdy project case, checking your work is relatively straightforward. Plug in an electric guitar and verify the output. Once this part is OK, switch off and add the components around the recorder input section. Check out the circuit again, this time feeding in a mono recorder input. Everything should transferred to the headphones. Finally, add the balance of the components. Adding capacitor C6 will shoot up the gain.

Test Setup

Plug in an electric guitar to J1, headphones to J2, and a recorder input to J3. Switch on the power, and turn up the guitar's volume control; most likely you will need to have your guitar's volume fully turned up. Use the electric guitar pacer volume control VR1 to vary the guitar signal through the headphones. Turn the recorder on and vary VR2 to suit you. *Note:* Keep the recorder volume control low to begin with, since the signal output from the recorder is much higher than that from the guitar—you don't want to deafen your ears. By balancing VR1 and VR2, you can now play along with your favorite music tracks. The combined sound mixing effect is superb, because the headphones isolate background noise and really make it appear that you're backing Cream's farewell concert at the Albert Hall as Eric Clapton's rips through his trademark thundering extended solos or in a quieter mood, you might be right up close to Lorrie Morgan as she sings her intimate "Something in Red."

Index

A

ac signals, projects outlined in book use, 36
Active components, 17
Amplification, most common requirement in electronics, 4
Amplifier combo, pre-amplifier power amplifier, 118–22
Amplifier with high-frequency cut filter, high gain inverting, 98–102
  circuit description, 99–100
  construction tips, 101
  parts list, 101
  test setup, 102
Amplifiers; *See also* Power amplifiers
  audio, 79
  audio power, 43–46
  audio system, 19–20
  class A, 44
  class B, 44
  LM 386 audio power, 20
  power, 107–10
    circuit description, 107–8
    parts list, 108–9, 109–10
    test setup, 110
  transistor-based power, 43
  Y, 68
Analog ICs (integrated circuits), 19–20
Assembly platforms, 26–27
Audio amplifiers; *See also* Audio pre-amplifiers
  building from discrete components, 79
  building from ICs, 79
Audio frequency band, 35
Audio power amplifiers, 43–46
  LM 386, 20
Audio pre-amplifiers, 35–41
  choosing safe ranges of component values, 41
Audio signal generator, 110–14
  circuit description, 111–13
  construction tips, 113–14
  parts list, 113
  test setup, 114
Audio signals
  electronic devices emitting, 35
  range of, 35
Audio system hookups, 71–78
  potentiometers, 76–78
  testing power amplifiers with piano-key style tape players, 72–73
  testing power amplifiers with record players, 74–75
  testing power amplifiers with Walkman-style tape player, 73–74
  testing pre-amplifiers, 75–76
  testing speakers, 71–72
Audio test set, 127–30
  circuit description, 127–28
  construction tips, 129–30
  parts list, 128–29
  test setup, 130

B

Batteries
measuring 9-volt, 62
9-volt, 32, 80
Battery snaps, 9-volt, 31–32
Beginning Electronics Through Projects (Singmin), 6, 33
Bias voltage, split, 8
Buffer
high-pass filters, 53
inverting, 92–95
circuit description, 93–94
construction tips, 94–95
parts list, 94
test setup, 95
low-pass filters, 53–54
non-inverting, 95–98
circuit description, 96–97
construction tips, 98
parts list, 97
test setup, 98

C

Capacitance, change in, 49
Capacitance values, reading, 13–15
Capacitive reactance, 40
Capacitors, 12–16, 38
coupling, 16
defined, 12
electrolytic, 13
need for, 13
polarity sensitive, 13
radial lead, 15
rules for connecting, 15
types based on physical structure, 13
variable, 15
Circuit boards; *See also* Printed circuit boards, 33
Circuit schematics, 4–6, 55–59
Circuits
flashlight, 55
frequency response of, 47
LED, 13
projects using LED, 29–33
transistor-based, 18
Class
A amplifiers, 44
B amplifiers, 44
Color codes, resistor, 9
Component values, choosing safe ranges of, 41
Components
active, 17
building block, 7–17
capacitors, 12–16
diodes, 16–17
resistors, 7–12
transistors, 17–19
electronic, 7–27
miscellaneous, 19–27
assembly platforms, 26–27
IC (integrated circuit) sockets, 26
ICs (integrated circuits), 19–20
jack plugs, 23–25
jack sockets, 23–25
LEDs (light emitting diodes), 25–26
switches, 20–23
passive, 17–19
Construction projects, 79–140
audio signal generator, 110–14
audio test set, 127–30
electric guitar pacer, 137–40
guitar fuzz pre-amplifier, 134–36
guitar pre-amp and buffer, 130–33
high gain inverting amplifier with high-frequency cut filter, 98–102
ICs used are LM741s, 80
inverting buffer, 92–95
inverting op-amp with gain x 10, 80–84
looking at portions of circuit schematics, 80
microphone test set, 122–27
non-inverting buffer, 95–98
non-inverting op-amp with gain x 10, 84–88
power amp with gain and bass boost, 107–10
pre-amp with bass-treble control, 102–6
pre-amplifier/power amplifier combo, 118–22
two-input mixer pre-amplifier, 114–18
variable gain inverting op-amp, 88–92
Contacts, ganged, 22
Coupling capacitors, 16
Current, measuring, 63

D

db (decibel) defined, 50
Digital ICs (integrated circuits), 19–20
Digital multimeters, making resistance measurements with, 64
DIL (dual-in-line) package, 8, 37
Diodes, 16–17
  forward-biased mode, 16
  polarity sensitive, 16
  reverse-biased mode, 16
  serves as protective devices, 17
DPDT (double pole, double throw) switches, 22
DPST (double pole, single throw) switches, 22
Dual-in-line package, 8, 37

E

Electric guitar pacer, 137–40
  circuit description, 138–39
  construction tips, 140
  parts list, 139
  test setup, 140
Electrolytic capacitors, 13
Electronic components, 7–27

F

Feedback resistors, 37
Filters
  buffer high-pass, 53
  buffer low-pass, 53–54
  high-pass, 51–53
  low-pass, 49–51
  simple designs, 47–54
  understanding, 47–54
Flashlight circuits, 55
Frequency point, upper cutoff, 50
Frequency response of circuits, 47
Function generators, 66–67

G

Gain, 36
Gain settings, 37, 45–46
Ganged contacts, 22
Generators
  audio signal, 110–14
    circuit description, 111–13
    construction tips, 113–14
    parts list, 113
    test setup, 114
  function, 66–67
  signal, 65–66
Guitar fuzz pre-amplifier, 134–36
  circuit description, 134–35
  construction tips, 136
  parts list, 135–36
  test setup, 136
Guitar pacer, electric, 137–40
  circuit description, 138–39
  construction tips, 140
  parts list, 139
  test setup, 140
Guitar pre-amp and buffer, 130–33
  circuit description, 131–32
  construction tips, 133
  parts list, 132–33
  test setup, 133

H

Half-supply voltage bias point, 38
High-pass filters, 51–53
Hookups, audio system, 71–78
  potentiometers, 76–78
  testing
    power amplifiers with piano-key style tape players, 72–73
    power amplifiers with record players, 74–75
    power amplifiers with Walk-man-style tape player, 73–74
    pre-amplifiers, 75–76
    speakers, 71–72

I

ICs (integrated circuits), 56–57
  analog, 19–20
  based designs, 79
  categories of, 19
  digital, 19–20
  op-amps, 36
  projects, 18
  reasons to use, 18
  sockets, 26
  special function, 20
Impedance, 38, 92
  input, 92
  output, 92
Instruments, test, 61–69

Inverting amplifier with high-frequency cut filter, 98–102
  circuit description, 99–100
  construction tips, 101
  parts list, 101
  test setup, 102
Inverting buffer, 92–95
  circuit description, 93–94
  construction tips, 94–95
  parts list, 94
  test setup, 95
Inverting op-amps
  with gain x 10
    circuit description, 81–83
    construction tips, 83
    parts list, 83
    test setup, 84
  variable gain, 88–92
    calibrating potentiometer, 90
    circuit description, 88–90
    construction tips, 91
    parts list, 90–91
    test setup, 91–92

J

Jack
  plugs, 23–25
  sockets, 23–25

L

Leads, polarity of test, 62
LED circuits, 13
  polarized devices, 30
  projects using, 29–33
    circuit boards, 33
    9-volt batteries, 32
    9-volt battery snaps, 31–32
    resistors, 32
    switches, 33
LEDs (light emitting diodes), 8, 25–26, 30–31, 56
LM 386 audio power amplifiers, 19, 20, 44
LM 741 general purpose op-amps, 19–20, 36, 37, 80
Loads, 44, 45
Low-pass filters, 49–51
  buffer, 53–54
Low-resistance loads, 45

M

Measurements with digital multimeters, making, 64
Meters; *See* Multimeters
Microphone test set, 122–27
  circuit description, 123–25
  construction tips, 126
  parts list, 125–26
  test setup, 126–27
Mid-supply voltage bias point, 38
Mid-voltage point, 38
Mixer pre-amplifier, two-input, 114–18
  circuit description, 116–17
  construction tips, 118
  parts list, 117–18
  test setup, 118
Multimeters, 9, 61–65

N

National Semiconductor, 44
Negative supply, 2
9-volt batteries, 32, 62, 80
9-volt battery snaps, 31–32
Non-inverting buffer, 95–98
  circuit description, 96–97
  construction tips, 98
  parts list, 97
  test setup, 98
Non-inverting mode pre-amplifier, 39–41
Non-inverting op-amp with gain x 10, 84–88
  circuit description, 85–86
  construction tips, 87
  parts list, 87
  test setup, 87–88

O

Octave defined, 50
Op-amps
  gain, 37
  IC, 36
  inverting, 80–84
    circuit description, 81–83
    construction tips, 83
    parts list, 83
    test setup, 84
  non-inverting, 84–88
    circuit description, 85–86

construction tips, 87
parts list, 87
test setup, 87–88
variable gain inverting, 88–92
calibrating potentiometer, 90
circuit description, 88–90
construction tips, 91
parts list, 90–91
test setup, 91–92
Oscilloscopes, 67–69
Output impedance, 92

P

Passive components, 17–19
PCBs (printed circuit boards); *See* Printed circuit boards
Platforms, assembly, 26–27
Plugs, jack, 23–25
Polarity of test leads, 62
Poles, 21, 23
Positive/negative supply, 2
Potentiometers, 8, 10, 76–78
calibrating, 90
Power amp with gain and bass boost, 107–10
circuit description, 107–8
construction tips, 109–10
parts list, 108–9
test setup, 110
Power amplifier combo, pre-amplifier, 118–22
Power amplifiers
audio, 43–46
LM 386 audio, 20
and stereos, 5–6
testing with
piano-key style tape players, 72–73
record players, 74–75
Walkman-style tape player, 73–74
transistor-based, 43
Power supply; *See* Supply
Pre-amplifier/power amplifier combo, 118–22
circuit description, 119–20
construction tips, 121
parts list, 120–21
test setup, 121–22
Pre-amplifiers
audio, 35–41
choosing safe ranges of component values, 41
with bass-treble control, 102–6
circuit description, 102–4
construction tips, 105–6
parts list, 104–5
test setup, 106
guitar, 130–33
circuit description, 131–32
construction tips, 133
parts list, 132–33
test setup, 133
guitar fuzz, 134–36
circuit description, 134–35
construction tips, 136
parts list, 135–36
test setup, 136
inverting mode, 36–39
non-inverting mode, 39–41
testing, 75–76
two-input mixer, 114–18
circuit description, 116–17
construction tips, 118
parts list, 117–18
test setup, 118
Pre-amps; *See* Pre-amplifiers
Printed circuit boards, 3
Projects
construction, 79–140
IC (integrated circuit), 18
using LED circuits, 29–33
Pulses, 69

R

Radial lead capacitors, 15
Reactance, 38
calculated, 48
capacitive, 40
defined, 40, 47
Record players, 74–75
Rectifiers, 17
Resistance measurements, making with digital multimeters, 64
Resistors, 7–12, 32
acting as potential divider circuits, 9
color codes, 9

Resistors, *Continued*
connected in parallel, 10–11
feedback, 37
measuring, 63
power rating of, 12
tolerance rating of, 11–12
values of, 10
variable, 8, 10
RF (radio frequency), 2
Rotary switches, 23

S
Schematics
circuit, 4–6, 55–59
complex, 5
defined, 55
describe electrical function, 5
introduction to, 5
Signal generators, 65–66
audio, 110–14
circuit description, 111–13
construction tips, 113–14
parts list, 113
test setup, 114
Signals
being loaded, 44
projects outlined in book use ac, 36
Snaps, 9-volt battery, 31–32
Sockets
IC (integrated circuit), 26
jack, 23–25
SPDT (single pole, double throw) switches, 21
Speakers, 44
from an impedance point of view, 45
testing, 71–72
Split bias voltage, 8
SPST (single pole, single throw) switches, 21
Stereos and power amplifiers, 5–6
Supply
positive/negative, 2
voltages, 2
Switches, 20–23, 33
configurations for, 20–21
DPDT (double pole, double throw), 22
DPST (double pole, single throw), 22
rotary, 23
SPDT (single pole, double throw), 21
SPST (single pole, single throw), 21

T
Tape players
piano-key style, 72–73
Walkman-style, 73–74
Test instruments, 61–69
function generators, 66–67
multimeters, 61–65
oscilloscopes, 67–69
signal generators, 65–66
Test leads
polarity of, 62
Testing
power amplifiers with
piano-key style tape players, 72–73
record players, 74–75
Walkman-style tape player, 73–74
pre-amplifiers, 75–76
speakers, 71–72
Throws, 21
Time base defined, 68
Transistor-based circuits, 18
Transistor-based power amplifiers, 43
Transistors, 17–19
active components, 17
come in huge variety of types, 18–19
defined as active devices, 18
NPN types, 18
PNP types, 18
Tubes, vacuum, 43

U
Upper cutoff frequency point, 50

V
Vacuum tubes, 43
Variable capacitors, 15
Variable resistors, 8, 10
Voltage bias point, mid or half supply, 38
Voltages
split bias voltage, 8
supply, 2

W
Ways, 23

Y
Y-amplifiers defined, 68